REFLEX CUBE AND THE NATURE OF THE UNIVERSE: HOW THE UNIVERSE WAS MADE

The space-time continuum explained in the language of a modern theorem; a theory of everything

By

Ford Van Hagen

REFLEX CUBE

AND THE NATURE OF THE UNIVERSE:

HOW THE UNIVERSE WAS MADE

Copyright © 2010 by Ford Van Hagen

Cover photo courtesy of www.Free-Stock-Photos.co.uk

ISBN 978-0-615-40265-9

Van Hagen Publishing   fvh@continuumspeak.com

CONTENTS

CHAPTER FOUR- REVIEW OF THE THEORY WORKBOOK

CHAPTER FIVE- THE MATH

Preface

We don't know if perfect geometric symmetry exists or not. It certainly doesn't exist in our reality, because perfection as an absolute is outside of the realm of human experience. This is why we can say with confidence, there is no such thing as a perfect circle or a straight line in reality. And since there is no evidence that mathematics has any existence beyond the realm of human thought, we have never had cause to explore the idea of perfect symmetry for any possible mathematics. One does not literally choose to go in the direction of this type of inquiry; rather, the currents of exploration will take you there after everything else has failed to produce results. I don't mean to diminish our accomplishments in mathematics. We have a wealth of mathematical information at our fingertips. But so far, none of our known mathematics has been able to explain the nature of the universe. This is why exploration and experimentation are so important in science. By using the simple geometry of a circle, I found a way to construct a theoretical model with perfect geometric symmetry and then learn from it, not knowing of course, if any of the information would be useful or not. As it turned out, I discovered a complete language with mathematics, a theorem I call "reflex cube". It didn't happen overnight. Imagine trying to decode an ancient message in hieroglyphics and not knowing very much about what you're doing. That's how difficult it was. The results, I'm sure most of you will agree, were well worth it. Reflex cube is the discovery of the mathematics of perfect symmetry. We're going to find out that a continuum is also a model of perfect symmetry, and that the nature of the universe is a continuum. We need to learn what a continuum is, and how a continuum works, before we can fully appreciate the mathematics of reflex cube, so good luck!

The book you are about to read and study was over twenty years in the making. The project began as a theory workbook, followed by hundreds of pages of notes, drawings, and a few more simple equations. When it became evident that there were going to be complicated mathematics involved, I decided to put the project on hold. Since I am not a mathematician, I was content to wait for the world of science to make the same discoveries and finish the new mathematics. So for ten of those years, the project remained idle. I reactivated the project after the turn of the century, when I realized the world of science was not catching up to my theory of gravity. Ten years is a long time to wait, and so I decided to try and finish the mathematics myself. And it worked. Every term in the new mathematical language has been carefully researched for accuracy in the definition. New words also had to be invented. I am not just proposing a new theory here. This is a complete science. It explains every question we have about the nature of the universe in a language of science and mathematics. The science is called "Continuum Dynamics Cosmology", and promises to be the foundation for New Era Three Physics.

If there is any dedication to be made, let this work be dedicated to the authors and professors I have studied. Let it be dedicated to the scientists who have labored their entire lives for the benefit of mankind. Let it be dedicated to all who have ever wondered about the nature of the universe, and to those who haven't, because the question was too big. Perhaps now they will be filled with wonder, because there will be an answer came true. Let it be.

Ford Van Hagen

Preliminary Q and A

Q- I don't know much about the universe, but I do know the universe started out with a big bang. Right?

A- Yes. Most people know about the big bang. I was a very young person when I first heard about it, and it sounded pretty scary. But after awhile, we were cracking jokes about who or what lit the fuse. Then later on we began to ask "what is the nature of the universe."

Q- What is meant by the word nature?

We're so used to referring to nature as "mother nature", that we immediately think of our natural environment on Earth. But when we ask about the nature of the universe, we are asking what is directly responsible for the whole universe, or what is the nature of all things? Just as mothers come before their children, so too does nature come before everything else. The science of cosmology is after this definition of nature to complete itself. That is its mission. Modern cosmology can solve every mystery in the universe, but if it can't explain what nature is, then it will have failed. It has to know what came before the universe and how it all got started. Now we simply ask, "what is nature".

Q-What mysteries in the universe are left to be solved?

A- They have narrowed down the particular questions to three. What is energy, gravity, and dimension. But it really is a two part question. What is time and space. They want to be able to identify the very first energy, as well as define the nature of energy. They want to be able to explain how gravity works in the universe, as well as define the nature of gravity. They also want to know if three-dimensional space has always existed the way it is, or if three-dimensional space had to be engineered like time into existence. So far they have only reasoned that space had to come first.

REFLEX CUBE AND THE
NATURE OF THE UNIVERSE

Q- Why does space have to come first?

A- Space had to come first because the big bang needed room for expression. It is natural for us to assume that space had to be there before anything else could be there. We just don't know how long it was there. Did it pre-exist, or was it included in the package when the universe of time began? The correct answer will help us to solve the mystery of nature.

Q- It's incredible to think that three-dimensional space could have been engineered like time into existence. And we were just getting used to the idea of the big bang. How are we ever going to find out about the question of space?

A- We need to solve the mystery of gravity first. Gravity holds the key to understanding everything else. It is the missing piece of the puzzle.

Q- Are we ever going to find out about gravity?

A- Yes, right here in this book. It's going to take a little bit of effort, but the study will be well worth it. Once we can explain gravity, space, and time, we will be able to explain the space-time continuum, and finish Einstein's work. Right now we have no definition for the space-time continuum, and we're not even sure what a continuum really is.

Q- Are we going to find out more about the continuum?

A- Yes, we are going to learn what a continuum is, and how a continuum works. We have the technology to explain it. We have always had the technology, but were unaware of the importance of its application. A circle gave us the invention of the wheel, and now it's going to give us the invention of a language that will help us to explain everything.

Q- Everything?

A-Yes everything, with the exception of just one thing. And you'll have to read the book to find out what that is.

Chapter One-
Continuum Questions

STRAIGHT TALK
ABOUT CONTINUUMS

We really don't know what a continuum is, or if a continuum exists. The word continuum is used quite frequently in science for a variety of reasons. We have given the word continuum a lot of definitions, but there is only one definition we are interested in, and it's the one we have no definition for in cosmology science. It's the space-time continuum definition. We simply don't know what a space-time continuum is, or where the universe came from.

Some say a continuum is something that "continues", while others disagree. The argument is, if something has to continue, then there must be separation in the continuum, when there shouldn't be any. We should expect a continuum to be in resolve as a completed process, so that there is no separation. We can use a circle to demonstrate each possibility; a continuum with separation, and without separation. The circle models each turn out to be excellent models for space and time. We're going to learn a theory called "circle of time theory", and it's going to be well worth the study.

The only clue we have about continuums is the

standard definition. The standard definition is a purely mathematical definition. This would be a great place to start looking around for answers, to the many questions we have about the nature of the universe. This is the kind of searching technology that makes cosmology exciting. It's called theory. Our predecessors were never lazy about it, and we're not going to be lazy about it either. Scientists, astrophysicists, mathematicians, theorists, particle physicists, quantum physicists, are the seekers of truth about "nature", and we owe a great deal of thanks to these people for their dedication and perseverance for the truths they have offered us.

WE NEED THE MATH TO TELL THE STORY

Almost everyone knows something about modern cosmology because it's everywhere in the media, all the time. Hence, modern cosmology has become a very popular science. Most everyone has heard about a mysterious thing called a singularity at the beginning of time, and that the scientists need a mathematical language to describe it. We would rather let the mathematics tell the story of nature, if the mathematics were known. But until then we can only provide a substitute language according to what we know, using known mathematics. These are called high tech theories, and they really do help us to stretch our imaginations, and to formalize our logic. We have become very skillful theoreticians and story tellers because of this. But nothing can compare with the actual discovery of a mathematical language telling the story about the nature of the universe.

Twenty years ago the importance of mathematics was not a priority. No one wanted to explain the "state of infinite density", a description Einstein used to explain what he could about the singularity, because nobody knew what it meant. Even today, the term infinite density lacks definition, and is being replaced by the term infinite dimension, which also lacks definition. And because it wasn't popular twenty years ago to insist that a mathematical language was going to be needed to explain the nature of the universe, theorists were free to theorize with their own imaginations about the early universe, and did not have to depend on the conventional rules of mathematics for inspiration. It was known however, that Einstein's energy equation was also the equation for creation, although the idea wasn't being popularized yet. People were beginning to see how relevant the "E" in the energy equation was in comparison to the big bang energy, and could easily see how energy becomes matter over time and distance, and how the equation when simplified can represent the whole history of the universe. And so for most theoreticians it became a matter of solving for "E", and not the direct question of what is the definition of infinite density. Solving for "E" meant further solution of the famous energy equation, as the question of "what is energy" was now in the spotlight as one of the most important questions in modern cosmology. It was reasoned that there had to be some kind of mechanics in play to account for the first energy, and probably existed in a similar fashion to the energy equation itself, showing forces, not numbers, in the equation. If E=MC squared represented the action of the universe, an equation for the

birth of the universe, or a "universe birth equation",
would have to represent nature, and solve for "E".

$$?=E$$

CONTINUUM PHYSICS

There is a science called continuum physics, which
specifically does not use Euclidian geometry, such as
points, lines, and circles, for equation purposes to describe
the universe. A form of algebra in three dimensions is
used instead to describe the workings of continuous
matter in infinite space and time. In this way, almost
everything known in the action of the universe can be
transcribed in the language of mathematics, including the
reaction of action. Often times in this science, the words
continuum and dynamics are used to describe the flow
mechanics of continuous matter. The word cosmology is
often associated with the word continuum, or the word
dynamics, but rarely if ever are the three words continuum
dynamics cosmology found conjoined in the same phrase
or sentence. Never before have these words been used
together to denote a new science. Continuum Dynamics
Cosmology is a new science in search of the space-time
continuum definition, and does use Euclidian geometry,
such as points, lines, and circles, for equation purposes to
describe the universe. Whereas the science of continuum
physics is a very accomplished science based on observation and measurement, and known mathematics, the
newer science is a branch of theoretical physics in search
of unknown mathematics.

A PHD IN DISINFORMATION

If you have a PhD in modern cosmology, which is not an exact science, then you don't mind having a degree in disinformation. The universe today is constantly being sold as a wild, weird, wonderful, and wacky place where anything goes. Empty space can push objects. Membranes in a sea of cosmic foam can encase whole universes. We're now supposed to have eleven dimensions of space, when spatial dimension itself has never been explained. Modern cosmology is open to a lot of speculation, because like the Wild West, it is uncharted territory. And since modern cosmology cannot always provide us with the truth, they are now in the business of providing us with possible answers. After a century of intense research and discovery, the three most important questions in modern cosmology were finally narrowed down to, what is energy, what is gravity, and what is dimension. And all within a generation or so, these questions have been completely abandoned, and replaced with high-tech mythologies, such as string theory, and membrane theory, to be taught in classes to the masses. Why did this happen? It wasn't out of disrespect for the many former colleagues who devoted their entire lives and careers to science. It was because we would have stalled in our efforts to navigate cosmology completely, had we dropped everything to focus on the three most important questions. After all, these questions are asking one big question: what is space-time? And we don't know. It's still a great mystery. But again, are we really trying to find out, or are we just looking for more possible answers to keep us in the business of cosmology? One thing's for sure, we're going to need something soon to keep us in business. Our high-tech mythologies have recently run out of popularity, after having been recognized for what they really are. Theories that raise more questions than answers are usually labeled as fiction, and headed for early retirement. What's next? And by the way, I believe it's time we stopped paying so much tribute to the ancient Greeks for their contributions to science. Modern cosmology should not be memorialized to a past civilization for the sake of defending our own modern high-tech mythologies.

As we begin to get back to the business of what is space and time, we need first and foremost, to review the logic of our classical definition of space-time. The definition tells us that dimension can be both non-physical and physical, because space-time is three dimensions of space, and one dimension of time. If time is one dimension of energy, then by definition, we should be permitted to reclassify the three subdivisions of energy into one classification as energy when discussing the forces of nature. Early on in cosmology, we are taught that the four fundamental forces of nature, electromagnetism, the strong nuclear force, the weak nuclear force, and the force of gravity, were all on equal footing at the beginning of time, and have since developed differently. This is an example of disinformation. Not all of these forces have developed differently, and there really are just two forces of nature, energy, and gravity.

There really are just two forces of nature.
ENERGY + GRAVITY

AVOIDING COMMON MISCONCEPTIONS

Modern cosmology is a jungle of information, much of it good, some of it bad. In the course of my research, I have learned it takes only a little bit of disinformation to throw us off completely. If I may make a suggestion to avoid misconception in the future, the term "fundamental forces" should be a term reserved exclusively for the astrophysicists, whose job it is to track the subdivisions of energy throughout the whole history of the universe. And by the way, I think they are doing a fantastic job of reconstructing the true story of cosmic evolution.

We also need to re-examine the term "space", and how we use it. Often times you hear the phrase "a universe of space-time", when in fact, the universe is a universe of time within space, not a universe "of space". When we refer to the term "universe" in this context, we should all be on the same page, and know we are speaking about the physical universe at large, an observable universe which is visible to the naked eye. We should also be aware that the physical universe is only occupying the "space". We need to view space as dimension, or "spatial dimension", and not as something intertwined with the physical universe. We need to break this old habit. We don't need to refer to outer space as the "ether", or use the term "fabric of space" anymore. Space is not spandex. It does not fold, spindle, or mutilate, or bend. What is in the space however, is very capable of curving, bending, and being influenced by other forces. So space contains the forces of energy and gravity, which are all around us. Some of it we can see, some of it we can't. The point is, space is very full of physical "stuff", and all of this stuff requires room for expression. Therefore, all forms of matter and energy are space dependent. We're going to find out that gravity is space dependent too. The subject of gravity, the other force of nature, will be discussed at length in the lessons ahead. When we learn about gravity, we will know why a revision in the general relativity theory is going to be needed.

When used in the proper context, the term space-time can be used to denote an area of real space with a specific location. For example, the space-time near the event horizon of a black hole is subject to extreme forces. These are called local space-times. The term space-time can also refer to the universe at large, having a specific location in real space. In the broader context, the physical universe as space-time itself is correct, allowing for the fact that an expanding universe beginning as a central point, has previously occupied all of the area of space we see today in one form or another. The big bang left a signature we refer to as the microwave background, and also sent out a tremendous shockwave which is still traveling well beyond the physical boundaries of the universe. There isn't one centimeter of virgin space in the entire universe left unaffected by something physical. And let us not forget about the many particles and waves in space today, and in the space the universe has occupied throughout its long history. Therefore, it is technically correct to refer to the physical universe as a space-time universe, because it couldn't possibly be anything else otherwise. In this instance of the broader context, the term space-time is still used to denote an area of real space with a specific location. In other words, a space-time universe is a local space-time. This is why we should refer to a space-time universe as a universe of time within space, and not as a universe of space-time.

NO ONE KNOWS HOW TO EXPLAIN GRAVITY

No one knows how gravity works. Go ahead and ask yourself. Ask anybody. Look up the definition. Get on the internet and try to find out. How does gravity work? What is gravity? Nobody knows. Many studies have been done, and many books have been written about gravitational attraction. The library is full of books about gravity. So why can't we define it? What is so mysterious about this force that we can't even begin to explain what it is, or how it works?

Let's say, for experimental purposes, that we can transport a Volkswagen back in time to the days of the caveman, and watch their reaction. How long would it take them to figure out what a car is, or to realize the principles of the combustion engine? No doubt it would take many generations to figure out the car's function as a transportation device, that is, if the car wasn't perceived of as a metal monster and destroyed first. Think of all the mind boggling minutes those poor slobs would spend trying to explain a car? Those cavemen, they're us! The car is gravity!

In modern cosmology, the question of gravity is at the top of the list, and rightfully so. Without it, we'd be floating around wondering why we didn't have it. Maybe then we would be able to explain it, or maybe not. In any case, it's going to be up to us to figure out gravity. The cavemen won't help us. They're long gone, and so is that Volkswagen. There is one place on the internet where you can learn about gravity, and how gravity works. It's my website, www.continuumspeak.com. The lessons provided on the website will teach you about gravity, and the space-time continuum.

NATURE AND ACTION

The study of cosmology is defined as the study of the nature of the universe. Many books have been written about modern cosmology, and are available to everyone for reading and study. These books inform us about the action of the universe and its history, but as yet, no one has been able to explain the nature of the universe in any meaningful way.

There are two parts to the mystery of the universe, nature and action. Nature has to do with what came before the universe, and the action begins with the birth of the universe. We are not going to figure out the nature of the universe, until we know everything we can about the action, and there are many questions left to be answered. We know the action begins with the appearance of a thermodynamic arrow of time in space, which has often been described by scientists as "something from nothing". If and when we can explain how this happened, nature will no longer be a mystery, and the science of modern cosmology will be complete.

Continuum dynamics cosmology is a two part

study which parallels the action of the universe, and the nature of the universe. The study of continuum dynamics in the logistical interpretation as circle of time theory is a study of the action, and the study of continuum dynamics in the mathematical language of a modern theorem is a study of nature. Circle of time theory is helping us to understand the action, so that the mathematics in nature can be understood. Circle of time theory has a wealth of technical information to offer, and this will help us in our quest to understand the mathematics of nature, but it is the mathematical language of a modern theorem that will ultimately complete the science of modern cosmology.

(Logistical interpretation- reducing mathematics to logic; the uncanny ability human beings possess which allows them to filter out all of the non-essential mathematical data in their daily routines; the ability to be specific so that an orderly progression of logic can proceed. However, when we do theory, we don't know if our logic is correct or not. The only hope for a theory is for the theory to be validated by mathematics. This is when a theory becomes a theorem. A theorem is the highest order of theory we can ever achieve).

SEARCHING FOR ANSWERS

The universe is still a huge mystery, and all of our latest popular theories have been labeled as fiction. If and when we discover the non-fictional version of reality, and the nature of the universe becomes known, we will have only discovered it, not invented it. Until then, we will have no choice but to continue to make up theories as we go along, hoping one day for a non-fictional account of nature, the real story, to surface. Meanwhile, we will keep ourselves busy by inventing more theories, and they too will be labeled as fiction, because that's what we do. We are experts on fiction. Nature is the only expert on reality.

We are allowed as theoretical observers to step back in time anywhere we want to go, even into the first second of the beginning of time. However, in order to preserve our field of observation, we are not allowed to go beyond the first nanosecond. It is reasoned that if we do, we lose our field of observation, the observation itself, and the observer. A boundary condition in science known as the Planck moment of creation became established after it was argued that empirical evidence can only be gathered through observation. It has been a long held belief that any chance for an observation beyond the Planck moment of creation is impossible. Today, scientists are rethinking this boundary condition. They are asking, what if there is information beyond this moment? Would information, if we could learn to decode it, qualify as an observation? It is believed that information does not require any real space to exist in, and we know that observers do require real space to exist in. So the answer depends on what space is, and we don't really know yet. That is why the question of "what is dimension" is one of the leading questions in modern cosmology today.

If you added up all of the collective time we have spent thinking about the universe and solving it's mysteries you would realize how far we've come in the past century. We know that energy and matter are not distinct as once thought. We also know that we have to explain the nature of the universe because it has a beginning. If we hadn't discovered the microwave background signature of the big bang, we would still have two theories to choose from, not knowing which one was correct, big bang theory, or steady state theory. One essentially says the physical universe has a beginning, is finite in size, and has nature. The other says the physical universe is infinite, has no beginning or end, and has no nature. Without nature to explain, the steady state theory is the easier of the two logical choices to explain. Einstein proposed in the steady state theory that the space-time continuum was the material universe for infinity into the past without beginning, and forward into the future without end. Einstein thought that by giving both space and time equivalency, he could show how space and time were interrelated, or interconnected. The same strategy had worked very well for energy and matter in the energy equation, which shows mass/energy equivalency. And as long as there is no question about the nature of the universe, the energy equation can stand alone and needs no further solution. The space-time continuum in the steady state theory is defined as three dimensions of infinite space, along with an infinite dimension of physical universe, and is the classical definition of space-time simply taken to infinity. Space-time is defined as three dimensions of space, height, width, and length, and one temporal of time. It is apparent that the easier of the two theories, steady state, is also the incorrect theory, and that further solution is required for the definitions of the space-time continuum, space-time, and the energy equation. Since the energy equation can be simplified as a big bang modeled equation, where energy becomes matter over time and distance, we now have to try and solve for "E" in the big bang. We need to know where the energy came from and how. We need to know what came before energy and equals "E". This is why "what is energy" is one of the leading questions in modern cosmology today.

(The evidence of the microwave background signature as proof of the big bang is being challenged by steady state theorists, who have suggested the signature may actually be that of low- temperature hydrogen. They still have their foot in the door, and so the truth becomes a matter of personal preference. If there was no big bang and the physical universe is infinite in all directions, then there really is nothing scientific or religious to discuss any further, as there is no nature to the universe. Still, there has never been a full explanation of what gravity is or how gravity works in the steady state theory. People who dismiss the importance of gravity are usually considered to be science illiterates; however there appears to be an exception to this rule. The opinion of a brilliant scientist should never be discounted. It is possible that the universe is a never ending ensemble of dead rocks floating in space, and is completely devoid of any purpose or scientific meaning, and so it really does become a matter of personal choice. Quantum uncertainty has already predicted that we can never be sure about anything, including the uncertainty principle).

The question of "what is gravity" still remains one of the leading questions in modern cosmology, because the mystery of gravity and how it works has never been solved by any scientist in the history of humankind. The centralized theme in the general relativity theory is that gravity is geometry. This means that gravity and geometry are unopposed; they are synonymous. Everything in the physical universe has a corresponding geometry called "shape", or "form", for which gravity serves as a compliment. What separates pure mathematical geometry from the opposite spectrum of physical geometry is only one thing- density. Mathematics is information, and doesn't need space to exist in, while physical geometries require space to express their form and densities. Physical geometry is dependent upon space for expression, and it is believed in the general relativity theory that any physical geometry large enough can curve the space around it. At the time, this was the only logical way to explain gravity, as something unopposed, something local. Gravity is either unopposed to geometry as geometry, or gravity is opposed to geometry as something else. We couldn't begin to explain it as counterforce before, but we can today. Only space can be unopposed. Time, energy and gravity, are opposed, and we now have an explanation for it. Is gravity geometry? The answer is no. Is gravity physical? The answer is no. Are there gravitons and gravitinos? The answer again is no. Whereas energy is a physical force of nature, gravity is a non-physical force of nature. There is a whole new science being built upon the question of gravity as counterforce; as pure momentum opposing energy. The science is called "Continuum Dynamics Cosmology", and dares to explore the unknown physics modern science cannot or will not explain.

No one has ever proposed gravity as counterforce to energy, because it doesn't seem logical. How can gravity be opposed to energy? You won't' be able to explain how gravity works if you cannot define what gravity is, or where it's coming from. So if gravity is opposed, what is it and where is it coming from? We are getting ready to explore this question, as we begin the study of continuum dynamics cosmology in the logistical interpretation as circle of time theory. We already know the action of gravity moves from concentric space, or surrounding space, to the centrosymmetric position of the physical geometry within that space, such as planets and stars. This arrangement of gravity provides a uniform distribution of force which holds everyone on the planet simultaneously, and is the reason why relativity theory believes the force of gravity to be an inherent property of all physical geometries, like the geometry beneath our feet, planet Earth. If gravity is unopposed as physical geometry, then the action can be explained as a pulling in action from the surrounding space by a physical object whose gravity wants to concentrate at its center. Gravity is much easier to explain as a local phenomenon, rather than as something non-local, however there has never been a full explanation of why gravity works in this way.

In the next chapter, we are going to demonstrate how the action of gravity works, by using the model of a circle. The use of geometry allows us to illustrate gravity in the form of a modeled equation. Gravity can never be illustrated as a pure equation like the energy equation, but it can be illustrated as a model of symmetrical expression. This allows us to describe gravity using the language of mathematics.

There are two things we should remember when trying to explain the nature of the universe: for every action there is an equal and opposite reaction, and there are only two forces out there, energy and gravity. Once we realize these forces as "action" and "equal opposite reaction", we can begin to visualize the mechanics of the universe. When we say time is opposed, we are saying energy and gravity are opposed. This means of course, that the big bang was synchronized with precision based mechanics, and that energy and gravity represent force and counterforce.

THE QUESTION OF UNKNOWN PHYSICS

Many scientists are in agreement that nature has unknown physics, because it is believed that gravity has precision based mechanics. Yet modern cosmology will go to great lengths not to explain how precision based mechanics can exist in nature. This is because the existence of unknown physics can't be proved either way, and gravity is still an unknown. The best we can do is to try and explain nature with theoretical physics and without physical evidence. This flies in the face of having a theory which can be tested and supported using the scientific methods of observation and experimentation. So the next best thing is mathematics. In theoretical physics, when you introduce mathematics into a theory, you can give the theory an edge. This gives the theoretician an opportunity to get around the question of precision based mechanics and the question of gravity all together.

One of our modern theories tries to get around the question of precision based mechanics by using the word infinity. The theory is called membrane theory. Given an infinite set of trial and error, nature can finally allow for a working universe to exist, regardless of how much precision is involved. The theory seems plausible at first, until you begin to realize the scope of the effort. That can be explained too if nature has an unlimited supply of energy and time. But is this a sound theory? Is this sound cosmology? It would be, if you could explain what infinity is, or how the universe works. Unfortunately, the theory doesn't explain either, and has only explained some of the effects in nature without explaining anything associated with cause. Which means we are still left with more questions than answers when you suggest an infinite set of trials to get a universe just right. For example; why does nature have to have so many failures before nature gets one right? Why is nature so wasteful? Why is nature so imperfect, and why is nature so determined? These questions remind us a little bit about ourselves, which gives the theory a certain appeal. The only difference between nature and us is that nature is forever and we're not. Nature isn't perfect either. It just has more time on its hands than we do, so it tries to make a universe, over and over and over and over. We're not the first ones to give nature a human face. The Greeks beat us to it. After a decade's run of popularity, modern cosmology finally admitted M theory to be nothing more than a high-tech mythology. It was then that the theory was officially labeled as fiction, And yes, we have a tendency to human-ize nature, but the reality is, we need a physical theory to explain things or we'll be stuck with the unknown physics. Membrane theory is a little of both. It can't be tested, but it can be visualized. They say in the theory that when membranes collide, they cause big bangs, and big bangs happen all the time, because there are an infinite number of membranes out there. Eventually, one of the collisions will produce a working universe. When you think about it, the logic of the theory is rather pat. Given an infinite number of bullets, even a blind man will hit the target, no matter where you hide it.

Proponents of M theory would like to remind us that the gravity question was addressed early on in an experiment with a small magnet and paper clip. It was concluded that all of earth's gravity was defeated when the magnet lifted the paper clip off the table. However, it wasn't reasoned that both the magnet and paper clip had to return to the surface of the table afterwards to remain at rest, according to the law of gravity. Nor was it reasoned that since gravity is still an unknown, you cannot use X the unknown as a quantitative sum to represent all of earth's gravity. The theory went ahead anyway to label gravity as a weak force. This lead to the conclusion that gravity was either leaking out of our universe or leaking in from an outside source. The outside source was identified as a possible membrane in a sea of membranes, and the innovative idea of the multiverse was born. M theory is considered remarkable because it has revolutionized our thinking. It opened up the possibility of time existing before the big bang, and was the first widely accepted theory to transcend the boundary condition of the Planck moment in a search for the unknown physics.

RELATIVITY IN NATURE

The theme of relativity has a much broader application in nature than even Einstein realized. Einstein gave us the theory of special relativity, and the theory of general relativity, and he was working on his third theory of the space-time continuum, which was never finished. He had always been convinced that space and time were somehow interconnected, but he never took the time to fully explore the continuum definition. Perhaps this would have been too simple a method. Or perhaps the complexity of the

REFLEX CUBE AND THE
NATURE OF THE UNIVERSE

subject would not allow a departure from the routine mathematics. Whatever the reason for the oversight, I believe Einstein would have eventually figured this out, and completed his theory of the space-time continuum. It would have meant a revision in the general relativity theory, but a trade off for the new knowledge would have been worth it. We are living Einstein's dream today, right here in this book. Einstein believed in a space-time continuum, even though he couldn't explain it as nature. Today, we can explain it, and prove Einstein was right. With a little bit of study, we can learn how to visualize the space-time continuum for the first time, and begin to appreciate how dynamic a continuum is, and how pro-found nature can be. As cosmologists, we have always wanted to be able to explain the nature of the universe with the proper mathematics, and now we can. It just took awhile to get the mathematics right. This is understandable once you begin to realize, that what is not possible in the action of the universe, is often times possible in nature.

And since we are on the subject of nature, now would be a good time to review the standard definition of the word relativity. Relativity is "a state of dependence in which the existence or significance of one entity is solely dependent on that of another". Perhaps this is why Einstein so strongly believed that space and time were somehow interrelated. Today, most cosmologists are in agreement that space, or spatial dimension as an entity, is not physical, and therefore cannot have physical bounda-ries. The logical conclusion to this realization is that space is infinite. The mathematicians however, cannot agree on a definition for infinity. It may seem humorous at first, thinking about a bunch of serious mathematicians sitting around a large table, arguing about infinity. But this is a good thing. If they all agreed, we would be stuck with one definition. Most everyone, including mathematicians, are in agreement about the logic of infinity, as something without beginning or end, and if space really is infinite, then there are no identifiable opposites in space as an entity. This is going to present a paradox in our view of space as a continuum. Who said learning the truth about the space-time continuum was going to be easy? It is time to put on your thinking caps. You may already be familiar with the lessons coming up next. They are as they appeared on my website, www.continuumspeak.com in September of 2008. The introduction begins with "continuumspeak is the only classroom of its kind where students are permitted to view the space-time continuum for the first time". I hope everyone, with a little bit of study, will get to enjoy the view.

Chapter Two-
The Lessons

A VIEW OF THE SPACE-TIME CONTINUUM: GRAVITY AS AN EQUATION

What is a continuum, and why is it important for us to explore the continuum definition? The universe is a very big place, and it's going to take something very, very big, perhaps on the order of a continuum, to explain it. Continuum dynamics (cosmology) is a new science in search of the space-time continuum definition. After all, the ultimate goal of modern cosmology is to finally explain the nature of the universe. Exploring the continuum definition has given us a new discovery: how to read gravity as an equation. When we ask, what is the nature of the universe, we should also ask, what is the nature of the space-time continuum.

The three leading questions in modern cosmology are, what is energy, what is gravity, and what is dimension. Over the years these questions have been refined to, what is the nature of energy; what was the first energy, what is the nature of gravity; how does gravity work, and what is dimension. Today we are going to learn how gravity works. In today's lesson we will be exploring the continuum definition, building a theoretical model for a continuum, finding an application to the model, and learning how to read gravity as an equation. That's right! Gravity can be illustrated as a model of symmetrical expression in the form of an equation, and you can learn how to read this equation in a very short time. The equation is based on a theoretical model for a continuum, so a quick study course in continuum dynamics is needed to show how gravity's application to the model works. In this theory (circle of time theory), we use the geometry of a circle as a model to show the dynamics of a continuum. The simple geometry of a circle can help to explain the nature of the universe, if the nature of the universe is a continuum.

According to the standard definition of a continuum, a continuum has to contain all of the rational and irrational numbers, or the set of all real and imaginary numbers including their negative images. By reducing these mathematics to logic, we can arrive at a logistical interpretation of the standard definition, that a continuum has to contain the opposites. When space (dimension) is viewed as infinite, and time (the universe) as finite, the logistical interpretation of the standard definition is fulfilled. Interestingly enough, a circle also fulfills the continuum definition. A circle contains both a beginning point and an ending point, that when connected, form a complete circle. Infinity also fulfills the continuum definition having no beginning and no end (or so we think). A circle with perfect geometric symmetry is like infinity; there are no reference points for beginning or end in the circumference of a circle with perfect symmetry, and this is why a perfect circle is a good model for a continuum.

REFLEX CUBE AND THE
NATURE OF THE UNIVERSE

Although perfect circles are never found in the action of the universe, we cannot discount the possibility of perfect symmetry existing elsewhere, in nature. There is nature, and then there is the action of nature, the universe.

(Lesson helper: We're looking for a way to describe a continuum, and a perfect circle is the best tool to work with. Why? Because a theoretical perfect circle is also a way of describing infinity, and a continuum should also represent infinity. Right now we're not worried about a perfect circle being only theoretical. We're faced with a new question. We can compare spatial dimension (real space) to infinity, and we can compare spatial dimension to a continuum, and we can compare spatial dimension to a perfect circle. But there is something missing. There are no reference points for beginning or ending. There are no opposites to compare with. One could argue that the opposite of spatial dimension is no spatial dimension. However, we can't visualize this opposite with any of our sensory awareness. We must be able to "see" the opposite in order to make a comparison, even in theory. Therefore, to qualify as an opposite, an opposite must be at the very least identifiable. So the new question we are faced with, is what is the identifiable opposite of spatial dimension?).

Circle of time theory is based on the rationale that a continuum cannot have separation, and must be interconnected, like a circle. Relativity means interconnectedness. A circle is balanced, is relative to itself, and has resolution as a completed process. If something has to be continued, then it is not a continuum without separation. According to the standard definition of a continuum, a continuum has to contain all of the rational and irrational numbers. A continuum has to contain the opposites. This becomes problematic for a circle with perfect symmetry, because without reference points for beginning and end, there are no identifiable opposites. If the same is true for a continuum without separation, then opposites are going to be needed to re-qualify a continuum without separation in the continuum definition. Space, a continuum without separation, cannot remain an isolated system and still remain in the continuum definition. Therefore, a continuum without separation has to contain its own opposite, a continuum with separation. So, where is the opposite? Where is the continuum with separation? This is called paradox. In order for a continuum to become a true circle of relativity, with identifiable opposites, a solution to paradox must be found. (See Lesson Two for relativity circle.)

Gravity is one of the biggest mysteries in the universe, besides the universe itself. We know a lot about energy because we can see it, but gravity has always been invisible to us, until now. Here is the equation: "Relativity's all of time in suspension model of the inverse squared without resolution". Here is the application: In my theory workbook, "Universe Birth Equation", I propose a theory that a matter universe and an anti-matter universe are created in the moment of the big bang, with the positive matter going in one direction, and the negative anti-matter going into the other, in an explosion/implosion scenario. The implosion is the force of gravity going inward into the inverse direction, becoming a sufficient force of resistance, enough to hold back the universe from running away with itself at escape velocity. The positive matter universe is still expanding today, while gravity keeps running at a smooth fixed rate. The question now becomes, why is gravity running at such a smooth fixed rate? The answer is "circle of time theory". The force of gravity goes all the way around its circle of time and meets up with the matter universe already in progress, and is blocked from resolving in its continuum. (Lesson helper: When gravity goes all the way around its circle of time, it is navigating in three dimensions, as opposed to two, and this can be confusing. Let me give you an example of how a continuum works in three dimensions. If I were to leave my chair and travel in a continuum, I would arrive back through the inverse direction, or from the inside out. The opposite is true in reverse. If gravity wants to go in the inverse direction, then it will reappear from concentric space, or outside in, to sit back in the chair). This accounts for the action of gravity throughout the universe in its relationship to matter. Gravity is counterforce, opposing matter, and is in square with matter. This is why gravity is holographic and realized only where matter is present. This explains why the action of gravity moves from concentric space to the centrosymmetric position of the physical geometry within that space. It explains why the force of gravity is inversely proportional in strength to the density of matter; Law of Inverse Square, and why gravity is running at such a smooth fixed rate, and why gravity is so fine-tuned. Gravity appears to have precision based mechanics and is not the direct result of the implosion, but rather a force that can achieve its circle of time almost instantly. The logistical interpretation of these dynamics is very limited because the physics are unknown to us. At least now we have a working theory of gravity and a closer look at the action of nature: two opposing arrows of time, a thermodynamic arrow, and an arrow of gravity; matter/energy opposing pure momentum.

Let's review the equation in full: The continuum without separation and without form represents "space", or open spatial dimension, as a continuum with resolution. In mathematical language this is called "inverse squared with resolution". The continuum with separation and with form is "Relativity's all of time in suspension model of the inverse squared without resolution". The continuums are contained, one in the other, in equilibrium, with form being the universe, and the universe having a function as the solution to paradox.

The gravity equation is the highlight of my 200 plus page theory workbook, "Universe Birth Equation", in the first edition, and remains an unpublished work. More mathematical language is needed to substantiate the theory and explain the unknown physics. The search continues in the second edition for a model of symmetrical expression in the form of an equation to represent the forces of nature, matter and anti-matter, in the birth of the universe- a universe birth equation. If such an equation is found, and we can explain the first energy and how energy was formed, then we should be able to explain continuum dynamics in the mathematical language of a modern theorem; a theory of everything.

The continuum in the theory has the appearance of a mathematical information system with operations capability, but in order for us to determine whether or not this appearance is actual in nature, the exact configuration of the system has to be identified and understood. Only then will we be able to explain how a continuum can perform the necessary operations for making a perfect universe. There is much evidence of mathematical implication in our studies of the universe to support the view that the universe is perfect. Scientists are now referring to the universe as "software". There is no evidence, however, that mathematics or mathematical information has an independent existence beyond the realm of human thought. Even so, we should not be surprised to learn that nature is indeed mathematical, and that continuums are real, continuums are dynamic; they make universes.

SUMMARY AND CONCLUSIONS

The theory of the space-time continuum is being revived by an amateur physicist who proposes gravity as counterforce, and the continuum as a dynamic process. The theory of the space-time continuum has never been fully explained because space and time have never been fully explained. If a continuum, by definition, has to contain the opposites, then space and time can easily qualify in the continuum definition as opposites in a space-time continuum. A comparative study analysis begins with the classical definition of space-time. Space-time is three dimensions of space, height, width, and length, and one temporal of time. This means that space is infinite, and time, the universe, is finite. Three-dimensional space is non-physical, without boundary conditions, and is formless. Time, the universe, is physical, has boundary conditions, and has form. Space is three superimposed dimensions as one. Time is one dimension within three. Space is unopposed. Time is opposed. Time is a universe of energy and gravity. If time is opposed, then energy and gravity are opposed. Conclusion: gravity is counterforce. There is a whole new science being built upon the question of gravity as counterforce, as pure momentum opposing energy. The science is called Continuum Dynamics Cosmology, and dares to explore the unknown physics modern science cannot, or will not explain. Only when we look at space and time as opposites in a space-time continuum can we begin to explain the nature of the universe.

THE STORY OF GRAVITY IN SIMPLIFIED LANGUAGE

Gravity and energy were born in the big bang. This much we know. But they don't go in the same direction. They go in opposite directions. Although we cannot explain the nature of these two forces yet, we can follow these forces from the moment they appear, and explain them scientifically as they are today. We've been able to explain energy very well over the past century, partly because energy is extremely visible. Gravity on the other hand, is an invisible force, and so it's taken us a lot longer to explain it. How these two forces interact today was established in the earliest moments of time. As forces go, these two forces are very much opposites. When matter and anti-matter separated in the big bang, they became energy and gravity, respectively.

Energy goes outward, and gravity goes inward. Energy is forceful, chaotic, and slow. Gravity is smooth, uniform and very fast. Energy is very physical, gravity is not. Gravity is like the force you get when you open a vacuum sealed container and air rushes in. Gravity is pure momentum. Energy is the explosion, and gravity is the implosion.

What happens next is beyond our known physics. Somehow gravity is able to reappear all the way around three dimensional space to begin pushing against energy. Gravity becomes counterforce. Now energy and gravity are opposed. This all happens very quickly, while energy is just getting started. This assures a relationship of symmetry between the two forces. In this moment gravity is trying to resolve at its starting place, but can't find its beginning. Energy already occupies this space, and is unwilling to share any space with gravity. As a rule, all matter occupies its own space. Now the early forces are symmetrical, with the action of gravity moving from surrounding space toward the center of the first energy. This is why today, the action of gravity moves from concentric space, or surrounding space, to the centrosymmetric position of the physical geometry within that space (i.e. planets, stars).

Since the beginning of time energy has been dividing itself, and every time energy divides, gravity has to divide along with it. In this way, the force of gravity is holographic, and realized only where matter is present. The story of gravity can also be explained in scientific language. Gravity can now be illustrated as a model of symmetrical expression in the form of an equation. The discovery of the equation allows us to view space and time in the new definition as opposites in a space-time continuum, with space (three-dimensional space), as unopposed, and time (energy and gravity), as opposed.

LESSON TWO: CIRCLE OF TIME TECHNICAL THEORY- THE EQUATION IN REVIEW

A circle depends on the act of formation for its very existence, and it is the act of formation which also predetermines the circle's imperfection. This is because time does not allow for the perfection of symmetry in the formation of physical geometries. The fate of all geometric symmetry in the universe is subject to inaccuracies imposed by the physical medium, which is why there is no such thing as a perfect circle or a straight line in reality.

REALITY CIRCLE

When we draw a two-dimensional circle, the circle has to have a beginning point and an ending point, that when connected, form a complete circle. Once connected, the circle is said to have "resolution" as a completed process, or to be "without separation". However, the reference points for beginning and end in the circle will forever remain identifiable, because a seamless connection of these points is never allowed in reality. A reality circle is "opposed" because it has reference points for beginning and end. The circle is also said to "have separation" until these points are connected.

39

THEORETICAL CIRCLE

Circle of time theory predicts that nature is like an already made circle with perfect geometric symmetry. Theoretically, a perfect circle equals infinity, because there are no reference points for beginning or end in the circumference of a circle with perfect symmetry. A theoretical perfect circle is "unopposed", because it has no reference points for beginning or end, has always been in resolution, and can never have separation.

RELATIVITY CIRCLE

These are our best models for circles in both reality and theory. Neither circle alone fulfills the continuum definition. Only a relativity circle can fulfill the continuum definition. The sum of these two opposite circles, opposed and unopposed, is a relativity circle.
"RELATIVITY'S ALL OF TIME..."

If three-dimensional space is a model of perfect symmetry, then it is like a perfect geometric circle, but without the physical boundary conditions of a circle. The reason three-dimensional space is believed to be infinite, and without parameters, is because spatial dimension is not physical, and therefore cannot have physical boundaries. If nature is space, and a model of perfect symmetry, then space cannot remain an isolated system and still remain in the continuum definition. Space, a continuum without separation, has to contain its own opposite, time, a continuum with separation. Space is unopposed. Time is opposed, and since time cannot be opposed to space, time

has to be opposed to itself. This is why time has separation.

TIME:

Time has separation…

<MATTER/ANTIMATTER>

ENERGY><GRAVITY

…and is now opposed.

The two forces of nature are opposed, and because
they are opposed, they are limited. This is why the universe is said to exist in a state of finite suspension.

TEST QUESTION:
Each circle has a theme. Which thematic circle best
describes the universe as a model of finite suspension?
 A) Reality circle without separation
 B) Reality circle with separation
 C) Theoretical perfect circle
      The correct answer is B), Reality circle with separa-
tion. The circumference of the circle represents gravity's
circle of time, while the separation in the circle represents
the physical universe.
"RELATIVITY'S ALL OF TIME IN SUSPENSION…"

When we are exploring physical and non-physical systems in theoretical physics, we use two-dimensional models as "themes" to help us visualize what can only be conceptualized in three dimensions. A two-dimensional image can give us a visual experience, serve as a reference guide, and provide us with a synthesis of contextual information for the benefit of study. In the case of the equation, the configuration of a two-dimensional circle serves as an excellent model for visualization, reference, and synthesis, yet we still have to use our own imaginations when conceptualizing the study in three dimensions. (It should be noted here that we use a circle for convenience. A line can also be used to demonstrate a continuum, as long as the line is a-symmetrical, moving in one direction only, and comes around "full circle" to connect the reference points.)

"RELATIVITY'S ALL OF TIME IN SUSPENSION MODEL..."

IN REVIEW: HOW GRAVITY WORKS

According to our best scientists, gravity cannot have room for variance. If gravity were varied by a factor of one in 10 to the 40th power, the universe would not exist as we know it. We know the universe begins with energy, and that energy and matter are synonymous (energy equation). What we are learning in the theory is that anti-matter and gravity are also synonymous. We usually think of energy to matter as a slow process of transformation over time, however the process is immediate with anti-matter to gravity, so that gravity can square with energy as counterforce. Gravity is trying to resolve in its continuum, but instead squares with matter. The reality circle is incomplete. Time has separation and is now opposed. When gravity squares with matter, it also establishes the "action template". This is the action of gravity moving from concentric space, to the centrosymmetric position of the physical geometry within that space. If gravity could be seen like rain, it would be raining down symmetrically from concentric space, or three-dimensional surrounding space, toward a central point. The central point at this beginning moment of time is the "matter universe already in progress", or the "beginning energy", which still has to be very symmetrical to account for the precision based mechanics of gravity. This means that gravity has to be immediate in the navigation of its circle of time, perhaps nanoseconds, to insure fine-tuning. But gravity is not instantaneous. Anything instantaneous would not require time for expression, nor would it allow any time for an observer to make an observation. We know that time has to first begin before we can make theoretical observations about the early universe. Although we cannot identify and explain the first energy yet, we can say that location is everything. The beginning energy has position at the apex of gravity's circle of time, and since all matter occupies its own space, gravity will be in detraction of this space for the life of the universe.

What begins as the first physical geometry with an action template in the early universe is now, after the big bang, many billions of physical geometries with a corresponding number of action templates. The idea is similar to a hologram. A two-dimensional hologram is a three-dimensional image on a two-dimensional surface. When the image is separated into pieces, each piece retains the image of the whole; or each part of a hologram retains the whole information picture. Today, the action template is everywhere in holographic display throughout the universe wherever matter is present, and as long as there is matter in the universe, gravity cannot resolve. Now, when we look out at the stars, we can almost see gravity's action template at work, trying to resolve, in a holographic system called a "holo-matrix". Thanks to a synthesis of contextual information in the form of a modeled equation, we now have a new window of gravity, with which to view the space-time continuum for the first time in human history.

"RELATIVITY'S ALL OF TIME IN SUSPENSION MODEL OF THE INVERSE SQUARED WITHOUT RESOLUTION"

PARADOX REVIEW

Paradox simply says an opposite cannot stand alone as an isolated system, and that the theme of relativity, or interconnectedness, is also the theme of the opposites. The theme of relativity has a much broader application in the continuum definition. To remain in the continuum definition, space must get to its opposite of time, the universe. Modern cosmology is after an explanation for the sudden appearance of a thermodynamic arrow of time in space, and also wants a complete definition of space. Either three-dimensional space is a natural phenomenon, or three-dimensional space had to be engineered, like time, into existence. Space is a continuum without separation, and has no identifiable opposites. Therefore, space has to somehow get to its opposite of time to remain in the continuum definition. Time is a continuum with separation. Space and time together form a relativity continuum, to re-qualify in the continuum definition as identifiable opposite continuums. This is sound cosmology. The mathematicians have already reasoned that dimension, or space, had to come first, because the big bang needed room for expression.

I hope you have enjoyed the study of continuum dynamics (cosmology) in the logistical interpretation as circle of time theory. As we begin to see space and time in the new definition as opposites in a space-time continuum, we must also ask, how can nature be programmed in the continuum definition? Continuum Dynamics Cosmology promises to be the foundation for New Era Three Physics, if and when we can explain how nature is programmed in the continuum definition, as information, and why nature has to exist as information in the first place. The question is very similar to the question on everyone's mind in modern cosmology today. Why does the universe bother to exist at all?

The astrophysicists would like to know why the big bang didn't cancel out before it even got started. Usually, positive and negative in equal amounts represent a sum for cancellation. Some physicists view the slight imbalance of extra positive which made up the universe as hardly enough to make a difference one way or the other. They have since changed their minds, but it made them wonder if there exists in nature an even distribution of positive and negative for non-cancellation. This leads to the inescapable conclusion that nature has an unlimited supply of symmetric polarity, which is why some physicists have described the big bang as nothing more than a hiccup. Nevertheless, modern cosmology still has to provide a technical explanation for "uneven distribution", while quantum physics has the task of explaining "even distribution". The law of averages, for which all of the quantum theory is based on, has never been explained.

Chapter Three-
Continuum Answers

NATURE AS
INFORMATION

The idea of nature as information is nothing new. Proposed by the famous German physicist and mathematician David Hilbert, there should be a level of pure information in nature, which does not require any real space to exist in. It would be an abstract space, such as the space a computer uses to do millions of calculations per second. This kind of space doesn't exist as we know it, because it is only theoretical space. Today, some physicists call it "mind space". But to call it that is more of an understatement, and it may mislead us into thinking of the word "consciousness". There is no mind that I'm aware of that can do millions of calculations per second. Information space is simply a computational space where all information has existence as information. In the language of modern cosmology and quantum physics, information space is often referred to as "Hilbert space".

There is a way to test the validity of Hilbert space, although I doubt anyone has ever thought of it. Perhaps David Hilbert thought of it, but I don't know if he did or not. It's really just a quick test of logic. If zero as an absolute is all there is, then there is no information space, no Hilbert space. By the mere fact that we are here as observers, we must conclude that there is more to zero than zero as an absolute (not to be confused with "absolute zero" which refers to temperature and the absence of heat energy). It could be that this kind of logic is the kind a mathematician uses to come up with the idea of information space in the first place. According to what we have already learned about the continuum definition, if there is more to zero than zero as an absolute, then there should also exist an information zero. Nature would then be in a balance of having a zero for non-cancellation, and a zero for cancellation, in a continuum called a "zero function". This would also mean information and no information in a balance, and it would almost certainly mean the counter-existence of a "continuum function", to make the information space, or Hilbert space, a relativity continuum, with identifiable mathematical opposites. Cancellation zero cancels, leaving non-cancellation zero as no information, to balance with all information, infinity. And there's your Hilbert space.

Of course this is only deductive reasoning. The exact configuration of the system has to be identified and understood, before we can make sense of the continuum as a mathematical information system. The discovery of such a system will have to explain every question we have, save for one, according to the quantum theory of uncertainty.

THE CONTINUUM PARDOX EXPLAINED IN THE QUANTUM THEORY

We're trying to understand the continuum reality as perfect existence without the need for objective/subjective realities. This may be an independent reality without observations or observers. If Hilbert space is a perfect balance of mathematical opposites for non-cancellation, then Hilbert space, even though it may qualify as a relativity continuum, cannot remain an isolated system and still remain in the continuum definition. Hilbert space may have identifiable opposites, but no one around to identify them. The solution to this paradox may be a re-qualification in the fulfillment of a self-automated program at rest and ready to engage a universe as the next step. In the transition from no activation to activation, the continuum reality may simply be the equation moving from the non-expressive to the expressive, and without the need for a command. If this is correct, then nature is capable of opposite projection for maintenance in the balance of nature, and provides a counter-reality with observation and observers in the form of a universe.

Quantum theory predicts that once we get down to the final question about nature, and after answering all of the questions before it, we will not be able to choose the correct answer without direct observation of the truth. Direct observation violates uncertainty, and so the absolute truth will not be known. The only truth we will be able to know is that both possible answers of either/or are correct, and that reality is a superposition of states. Either the universe is an exercise in the continuum function, or a command is required to get it going, or both. Paradox has a multi-dimensional theme in both action and nature as questions and answers in a process of elimination, and so it should be of no surprise that paradox reaches conclusion in the final analysis as a superposition of states.

As observers, our reality depends on the act of observation in real time to assimilate the information of the observation. Conversely, what we're not looking at has no existence at all as information. The whole idea of observer dependent reality is to pinpoint the definition of time as the time present. There is no future because the future is a non-observable, and there is no past because the past is a non-observable. The only observable is the present moment and this is defined as observer dependent reality. That's why nothing exists unless you are looking at it. This does not mean all physical reality disappears outside of your field of observation. It means that everything else aside from the observation disappears as information, and only that which is being observed has existence as information.

It may turn out to be true that energy and matter are actually created by pure information. But for now, energy and matter, and space, apart from human observers, exists only in the objective reality. And we, the observers in the subjective reality, are dependent upon this objective reality for our existence.

Observer dependent reality goes a step further by separating consciousness as the observer, from the world of particles and waves as the observation, in an exclusive two-dimensional reality. The premise is that observer and observation are the only two truths. Some will argue that observer dependent reality goes too far in its reductionism of all things to only that which is observable, but the theory is technically correct. Matter is the only thing that qualifies as an observable, and all forms of matter can be reduced to particles and waves. This means that solid matter is an illusion created by consciousness. Of course this means we are the objective reality, and the world of particles and waves is the subjective reality. Again, there are two choices to explain the mechanics of form and the higher geometries we perceive as reality. Either consciousness is responsible for the illusion of form and matter, or nature is self-organizing and fully capable of performing multiple realities.

If consciousness is capable of formation and transformation, then the theory clearly establishes the role of observer as cause, and observation as effect. The continuum theory takes over from here to remind us about our dependency on open space, and about the continuum definition, and to provide us with the opposite of cause and effect in open space, which is no observers or observations in closed space. The continuum theory and observer dependent reality may disagree on whether the active roles of objective and subjective realities should be reversed or not, but the conclusion is always going to be the same; the opposites of cause and effect, of observer and observation, are only going to be active in open space.

COSMOLOGY PERSPECTIVE

A good cosmologist can see the value in the quantum theory. A good cosmologist can also find disagreement. The main contention is gravity as a non-observable. We can now explain the definition of gravity, a definition which has mystified us for ages, with a fair amount of accuracy. So, it's not as if gravity were completely invisible anymore, it's partially visible, right? No, gravity is a non-observable. Gravity, even quantum gravity, is usually discussed in the science of cosmology, not quantum physics. In the new definition, gravity is explained as the method of transformation from particles and waves to solid matter, in a slow process of compactification over time and distance. But we already knew that. We just didn't know how gravity worked, and now we have a logical explanation. We now know that to get the desired effect of solid matter, nature has to provide the opposites of force and counterforce in just the right proportions, so that a universe can exist in a state of finite suspension. And since finite suspension is only temporary, and time does not allow for the perfection in the formation of physical geometries, all of the symmetry in the universe will eventually be reduced to the particle/wave medium from whence it came. This is a medium which is made to insure that all things physical and imperfect will also experience entropy and decay, to the point where the original symmetry is unrecognizable. This will more than likely be the fate of the physical universe. But you never know. Who is to say the universe isn't capable of recycling itself again, and to be even more spectacular than it ever was before.

The reason we have been unable to provide a unified theory is because of the discrepancy between quantum mechanics and gravity. Quantum mechanics is no mystery. The science has given us a wealth of technological achievement, while the mechanics of gravity have remained virtually unknown. The mathematics of a modern theorem will inevitably solve any discrepancies between the two sciences of quantum physics and relativity physics.

SUPERPOSITION

The word superposition has many more applications in theory than in reality. This is because we live in a physical world where all matter occupies its own space. Every bit of matter has its own identity, just as every number has its own identity. Yet, in the world of mathematics, it becomes easy to superimpose numbers when doing mathematical operations. You can't do the same thing with matter. This is because matter has position. All things physical have position. Only that which is non-physical can have superposition, such as three-dimensional space. So space has superposition, and time physical has position. This is a new way of looking at reality. When we look up at our beautiful sky on a sunny day, we can say, and understand, that we are in the position, within the superposition of spatial dimension. So, if time is the position, and space is the superposition, what is the super?

Not knowing anything that you have learned so

far, this may seem like a difficult question. It was for me, and the question of superposition was also difficult. Now it seems easy, but you have to remember, I am a well seasoned theoretician. There is no way to break the news gently, that reality may indeed be a superposition of states. Now we're looking at the possibility of "no space", or the "super", as being real. This has to be the domain of the continuum, and of course, we already know that information does not require any real space to exist in. We know mathematics is information, and that mathematical operations are examples of superposition. Now that we know superposition is possible because we have space and time, the idea of Hilbert space, or theoretical space, existing in a superposition of states as a super reality, becomes a more real possibility. While the super's domain of closed space may yet be another opposite of three-dimensional open space, closed space is another continuum without separation and without identifiable opposites. Even if we could point to these two opposite continuums of open and closed space as identifiable opposite continuums, how then are you going to explain a universe when there shouldn't be any? In today's cosmology we pretty much have this figured out. We're aware of not only time coming out of the big bang, but we are aware of the possibility of both space and time coming out of the big bang, with space coming first, to allow a universe of time to have room for expression. We know that nature has to be more than just a double reality of open and closed space opposites, because both continuums are without separation, and therefore alike. Opposites attract, likes repel, and time has separation.

We should also remember that zero as an absolute and information zero are slightly different in the same way. They are the same in that they don't require any real space to exist in, yet they differ, because information zero does require theoretical space to exist in, while zero as an absolute represents complete and total cancellation. One could say at this point, that nature is an equilibrium of information and no information, but that still doesn't explain the big question of how nature is programmed in the continuum definition, and why nature has to exist as information in the first place.

Superposition is not as simple a concept to grasp as it seems. It's almost worth being an investigative science by itself. The theme of superposition can be found everywhere in both quantum physics and relativity physics, and plays a role in the duality of nature found in each of these sciences. We are familiar with the term "superposition of states" in the quantum theory, and in particle physics, the term is used to describe many of the characteristics of the electron. In the theory workbook, "Universe Birth Equation", the superposition becomes a destination for all matter/energy consumed inside a black hole. This would be a place where matter/energy is in a superposition of states between existence and non-existence. But it isn't until much later in these investigations that the meaning of "superposition" reaches its full potential. By breaking the word superposition down to its constituents of super, superposition, and position, we can begin to make a comparative study analysis of space and time, and then begin to apply this logic to the space-time continuum definition.

Perhaps one of the best examples of the word superposition is in reality. Every time you see a reflection of something in a mirror, glass, water, or any other reflective surface, you are seeing something in superposition. A reflected image is a reversed image in superposition of something physical with position.

MATHEMATICAL IMPLICATION

When we study the logic of the universe, we find there are areas which stand out to be highly coordinated and suggestive of a system. These areas, which imply mathematics at work, can only be interpreted as connotations, and are often referred to as symmetry, or coincidence. There are three highlights of mathematical implication in the history of the universe with the most obvious symmetrical distributions- 1) the big bang, 2) stars, 3) and gravity, each providing an example of what appears to be precision based mechanics. The highlights of mathematical implication are as follows: 1) only a small fraction of the big bang became our material universe and the big bang lasted a long time before it was finished. It would have been a pure radiation event for its entire life span if it weren't for the first several minutes of stable matter production, allowing our universe to form. This is when electrons and positrons were fusing together instead of annihilating each other into radiation (which is an exception to the rule). The survival ratio of these special "virtual pairs" was one to a billion over the entire length of the big bang, which after ten-thousand years resembled an opaque fireball, and after two-hundred thousand years, a glowing ember. The electron positron pairs that did survive the fate of radiation went on to become protons and neutrons, which later became stars and planets, and eventually us. Scientists call it an imbalance in the laws of nature which allowed for the survival ratio of one in a billion electron positron pairs, citing that had the distribution of positive and negative been any different in the big bang, we would have had a failed universe of pure radiation. Modern cosmology still has to explain the uneven distribution in the big bang, or the imbalance, which guaranteed a successful universe. We now know the very beginning of time is literally the birth of the universe, a universe which has outlived all of the radiation that followed in the big bang, save for the signature it left behind- microwave background. We know the age of the universe, 13.7 billion years. We can now personally identify with the very earliest moments of creation, which means we have the entire history of the universe in each and every one of us. 2) We also know that stars have to maintain equal amounts of oxygen and carbon during their long life cycles in order to perform nucleosynthesis. This is a very important example of even distribution, and is the second example in the highlights of mathematical implication. 3) The third example, gravity, is so fine-tuned, it cannot have room for variance. If gravity were any different, by even a slight variation, we probably wouldn't be here having a discussion about it. The universe would not exist as we know it.

THE EXPERIMENTAL MODEL

When I began writing the Universe Birth Equation theory workbook twenty some years ago, I had no idea the work would develop into a theory workbook. I had become very familiar with the important questions in modern cosmology, after reading many books, and could not resist the urge to write about my own theories. As many students in this area of science will tell you, the urge to produce theory happens quite frequently. There is and has been no other science of its kind which allows so much freedom of thought. In all of my research, there was almost no mention of the word continuum, and so I became curious. I soon became fascinated by the mystery of the word continuum, which to me meant never ending, or infinite. I thought we might be asking the wrong questions in our popular science, rather than pursuing the more difficult questions we needed to face. Now I understand why the continuum definition was on the back burner in cosmology some years ago. We already had our hands full just trying to contend with the many questions about the universe at large, that we didn't need something larger to worry about.

It was early into the theory workbook that I began to explore the idea of what a continuum was like, and had to use a technique called "creative visualization" to come up with one. I started to imagine a continuum as something coming around full circle, and soon came up with my first experimental model. The model was such a learning experience and such a challenge, that it became the inspiration for "circle of time theory". As we shall see, the task of connecting two reference points to complete a two-dimensional circle is not as easy when working in three dimensions.

As one of many thought experiments about the early universe, this one really made sense. I asked myself, what would a point of light do in the dark void of space, if left alone and unchecked, and without the force of gravity around yet to possibly influence it? I reasoned that this point of light would begin to expand, uniformly, symmetr-

ically, outwardly in all directions, and then expand expo-
nentially, trying to attract to the only known thing in the
universe, itself. And as this light begins to fill up all of
dimension, it somehow manages to transcend its concentric boundaries, and then bonds with itself through the
inverse, or from the inside out, at its original starting place.
That's how it's done. That's how reference points are
connected when something comes around full circle in
three-dimensions. The initial reference point has to be
approached from the inverse direction, in order for a
connection to be made. It can't work in any other way. I
had just constructed my first model for a continuum, using
infinite space as my canvass, and light as paint.

Unbeknownst to me at the time, the experimental
model would do more than inspire me to produce more
theories. The model would later serve as the basis for the
gravity equation. The model for how gravity works is the
same as the experimental model, only in reverse.

THE FIRST EQUATIONS IN THE THEORY WORKBOOK

In the theory workbook, I use the experimental
model to tell an elaborate story of how a continuum of
light can come into existence as a circle of time-light. It is
in the action of this mythology that I begin to notice some
interesting mathematics. The equation, one equals infinity,
results when the light completes the circle, because the

first phase of light to complete the circle has to represent all of the phases of light in the circle. This is simply a matter of deductive reasoning. All points in the circle have to be the same upon completion of the circle, and equal to both one and infinity. Much later in the theory workbook, the equation, one equals infinity, surfaces again in the inverse squared with resolution. All mathematical points in three-dimensional space are points of continuum resolution. It's the same principle, and validates the idea that all points in a circle with perfect geometric symmetry share the same identity. Space has to be a continuum with resolution, which is the same as a continuum without separation. Every point in three-dimensional space can therefore be represented by the equation, one equals infinity, and all points in space share the same identity. These are much different mathematics than we are used too, but that's the fun of it. It's called discovery, but discovery of what? Does it mean that nature is capable of producing "identicals" in a universe where no two things are exactly alike? The answer is yes. All mathematical points in three-dimensional space are exactly alike.

The real significance of the time-light experiment is not apparent early on in the theory workbook. Theoretically, in order for the light in a continuum of time-light to achieve "resolution", it also has to achieve "continuum momentum". With continuum momentum, the light would have the ability to be instantaneous in its navigation of the continuum from start to finish. So the idea of continuum momentum, or something instantaneous, is not yet fully realized in the theory until later. We still have questions about the inverse squared with resolution and the nature of space. If space had to be engineered like time into existence, then it more than likely was instantaneous. Gravity would have become space too if it were also instantaneous, but gravity is only immediate. Therefore, gravity, a click slower, is in the expressive, and space is not. How this was achieved will be fully explained in the language of modern theorem. It suffices to say, for a continuum to produce infinite space is as easy as turning on a light switch.   .

To understand the difference between the expressive and the non-expressive, imagine two electric fans, side by side, one fan active in its rotation cycle, and the other at rest. It is sometimes possible for both fans to appear to be at rest, in the non-expressive. Only when the active fans rotation cycle begins to slow down do you begin to notice the difference. Space is like the fan active in its rotation cycle, but appearing to be at rest, in the non- expressive. When the rotation cycle is slowed, you would get something like gravity, in the expressive. Now we can appreciate space in the non-expressive for having no aspect of time to it, which is why space is the complete opposite of time in the expressive.

We're not necessarily looking at light as the component of the continuum in the experiment anymore, because the whole experiment is in the expressive. We're now looking at momentum as the component of the continuum, a continuum momentum which has the ability to be everywhere at once, in the non-expressive. But we need to get back to the time-light experiment to review the second equation in the theory workbook, an equation which comes as a natural consequence of trying to connect the two opposite ends of a continuum together as one for "resolution". The rule is going to be that the connection has to be seamless, and perfect, so that 1+1=1. In the theory workbook, the two opposite ends of a continuum each have to represent an inverse, one for beginning, and one for end. These inverses represent the absolute threshold of a continuum, and will have to be synchronized for the sum of these inverses, or 1+1 to equal one. This can only happen at continuum resolution. This is when the inverses are sharing the same rate of momentum, and are allowed to synchronize as one when the circle is completed. The science of continuum resolution is going to take us into the mathematical language of a modern theorem, but first we need to learn more about synchronization. Then I want you to take a brief tour of the theory workbook to familiarize yourself with some of the important language, concepts, and ideas, with applications in modern theorem.

SYNCHRONIZATION

We don't need the ability to construct a perfect circle in order to understand the mathematics of perfect symmetry. We know that the opposite ends of a perfect circle are seamlessly connected somehow, and exist somewhere in the circle. We know that a theoretical perfect circle represents infinity, and that all of the points in the circle are also equal to one, for the equation one=infinity. The second equation in the circle, 1+1=1, is going to give us our biggest clue about nature. Since there are no identifiable reference points for beginning or end in a perfect circle, each point in the circle has to be represented by the theoretical connection of 1+1 to equal one. This equation can also be read as 2=1. This is our seamless connection in every point for a perfect circle. The two theoretical opposite ends of a circle or a continuum can be represented in each and every point in the circle, and synchronized as one, if each and every point in the circle share the same momentum. In the case of a circle with perfect symmetry they do. The momentum is equal to infinity, and is called "continuum momentum". Continuum momentum assures that all points in the circle share the same identity through a process called "synchronization". Synchronization occurs when the opposite ends of a continuum can share the same momentum, and become synchronized as one for continuum resolution. This is when all mathematical points in the circle become points of continuum resolution. And this is only the first application of the equation $1+1=1$. The next application has to do with nature itself. When we try to figure out what was going on in nature to begin with, we can only imagine a state of nothingness. But if nature truly is nothing and nothing is all there is, then nothing also has to be everything, for the equation of $2=1$.

And if nature is in resolution as $1+1=1$, then nature is already in continuum resolution at zero. Hilbert space can use closed space as a means for synchronizing mathematical opposites by containing them in a mathematical fusion. So now we're looking at continuum momentum, or infinity, as being synchronized with no momentum, or zero, just like the two fans. It is possible these two opposites are sharing location as one in mathematical fusion, with the same synchronization. This is what the information in Hilbert space is supposed to look like. Closed space may simply be the mathematical fusion of these opposites for a perfect synchronization, and to get to open space, would require nothing short of "infusion". Whether the infusion goes in, or out, the result is going to be the same, and the result is going to be instantaneous, in the opposite inversion from mathematical fusion, or no real space, to an infusion of all real space. At first we are very interested as scientists to learn about the cause and effect relationship of this movement, from closed space to open space, until we begin to realize that anything instantaneous does not require time for expression. However, we are also beginning to understand the continuum as a system of opposite inversion, and we are very much interested in the mechanics of this process.

Let's look again at the two fans. The one on the left is active in the rotation cycle, and the one on the right is at rest. The one on the left represents continuum momentum, or infinity, and the one on the right represents no momentum, or zero. If nature is already in resolution as $1+1=1$, then the two opposite fans are synchronized as one. This is why they both appear to be at rest. The opposites are sharing location in mathematical fusion, and this is exactly what you will need to get a universe. It is easy to understand how momentum can be synchronized. Remember gravity? Gravity is an example of synchronized momentum. Momentum is the key to everything. It really is the first "space". To get to the equation of $1+1=1$, all nature has to do, is to synchronize the opposite ends of a continuum. And since continuum momentum is already locked in the fusion, and its rotation cycle is already complete, infinity is now back to zero and is at rest. The synchronization of zero and infinity is therefore accomplished easily, and naturally, in the mathematical fusion of closed space. This is called 2 in 1 mathematical fusion, and it only happens in Hilbert space. This is the only place where mathematical opposites can have equivalency.

Information space requires at least one dimension, according to classical physics. So there has to be a superposition of at least one dimension for Hilbert space to exist in. Information cannot exist in the fusion of closed space, or the non-dimensional. So there have to be two types of fusion in closed space. One is for information, and the other is for no information. One is mathematical fusion, the other is fusion. Apparently, nature is both. Why?

This is one of the ultimate questions in paradox. Why does nature exist as information? Why is there an equilibrium of information and no information? Why is nature programmed in the continuum definition? There happens to be a very logical answer. It is the ultimate secret of the continuum, and the knowledge of this secret will be fully explained in the science of continuum resolution. The secret is also mentioned in one of the earlier chapters of the theory workbook. See if you can find the clue. If you can guess what the secret of the continuum is, then you are right on track with modern theorem. It has to do with what happens at continuum resolution, and it is something very special. Ask yourself this question: what is the most unique thing in the universe?

We know that a continuum is about continuum momentum, synchronization, and opposite inversion. We are beginning to learn about the continuum as a mathematical information system. And yes, we are going to be able to identify the exact configuration of the system. There is a mathematical theorem for continuum dynamics, and there is a universe birth equation. And the best news is, the mathematics are not as difficult as the mathematics we're used to. Eventually everyone with basic math skills will be able to explain modern theorem.

You have probably learned something about the configuration of the continuum as a mathematical information system without realizing it yet. Open and closed space continuums are the same, because both continuums are without separation. They are also opposed as opposite continuums, and they are unopposed as independent continuums. Even though the exact configuration of the system has not been identified yet, we know the system has to be able to sustain multiple synchronizations to account for these multiple continuum realities. And while we may be learning something important about continuum dynamics, we should also be aware that we're going to need the mathematics later on to validate the logic of these conclusions.

As we begin to summarize these conclusions, we are getting very close to the definition of a continuum. A continuum is continuum momentum, or synchronized momentum, and the information in the synchronization tells us that a continuum has to be fast enough to be in two places at once, representing both opposites, and also has to do this three times over to account for opposites which are alike, opposed, and unopposed. We are not used to the idea of opposites being the same in the action of the universe. To us, this seems like a paradox. However we are not talking about opposites in the action of the universe. We are talking about nature, and in nature, things that are not possible in our reality, are possible in the continuum reality. Furthermore, it doesn't matter whether opposites are mathematical or not in nature, as long as they are true opposites. For example, instead of saying zero=infinity, we could say, nothing is everything.

INFORMATION AT OUR FINGERTIPS

How lucky we are to have all of this information at our fingertips. As I said before, we did not have the luxury of anyone telling us much about the nature of the universe two decades ago, nor did we have the luxury of today's internet. Going beyond the Planck moment was taboo back then, and so, as students of cosmology, we had no choice but to study the action of the universe. And there was plenty of action to study. In fact, there was too much science to learn and remember. That's why after reading several books, I'd had enough. I became inspired, and decided it was time to develop my own theories. Since my interest was the nature of the universe, and there wasn't much discussion about it, I began to think, and write. And it took nearly two hundred pages, and almost eighteen months, to get to the point where I began to question the nature of "dimension", or "three-dimensional space", instead of taking open space for granted as I always had. This is when I discovered the idea of closed space.

As it turned out, the idea of closed space had previously been mentioned in a book by Heinz Pagels I hadn't read yet. If I had read that book early on in my research, the Universe Birth Equation theory workbook might never have been written, and we would still be in the dark about the nature of the universe. I began my theoretical journey through space and time by assuming that open space was always open space, and so I never questioned it, until much later. If someone had suggested the idea of closed space to me sooner or even suggested that spatial dimension had to be engineered like time into existence, I most certainly would have dismissed either suggestion as next to impossible, and then headed for the window. And if someone suggested that nature could be described by a set of numerical equations, I would have then headed for the door. In the beginning of the theory workbook, I make it expressly known, that I'm looking for forces, not numbers, in any equations for the birth of the universe.

Twenty years ago, modern cosmology had narrowed down the two most important questions to what is energy, and what is gravity. We needed to explain the "thermodynamic arrow of time", a term which was being used to designate the big bang as the new definition of "time". Modern cosmology was going high-tech, but it would take another seven years for the question of dimension to finally enter into the arena, allowing the broader question of what is space- time to be ready to take center stage. I remember being thrilled when this happened. But I also remember asking myself, what took you guys so long?

The question of dimension was brought about by the mathematicians, who were in a search for the definition of the number one. They are the "mathematical" cosmologists, searching for the nature of numbers, so that they can apply mathematical logic to the universe. They wanted to identify the very first thing in existence. It was reasoned that the big bang needed room for expression, so that dimension had to come first. Spatial dimension, or three-dimensional space, was then assigned a cosmological value of one, as being the very first thing. It was a unanimous decision, and the question of what is dimension was ushered in as the third most important question in modern cosmology.

The question of the nature of energy was never solved for in the first edition of my theory workbook. After eighteen months of writing theory, I found myself back at the Planck moment, trying to re-explain the first energy with better science, but I knew it was hopeless. And so I had to close the work, feeling as though I had missed the target. I felt very confident however, that as an amateur physicist, I had provided a technical explanation for gravity, and I was looking forward to the day when the new information would have more of a story to tell. I knew I was scratching at the surface of a theorem, and not being a mathematician, realized I needed more time to think. I didn't know where to begin again. I told myself to be patient and take notes, and as I began to write, I felt as though I were the only person in the world who knew something about the nature of the universe, and I felt alone. That's when I rediscovered Hilbert space, and began to think in terms of mathematics. The possibility of working in theoretical space opened up a whole new universe for me. The loneliness passed quickly. I was soon busy experimenting with math, drawing diagrams, and making new discoveries.

A WORK OF EXPLORATION

The "UNIVERSE BIRTH EQUATION" theory workbook is a work of exploration. It begins with a journey to the early universe for theoretical observations, and soon becomes a quest for the meaning of continuums. From there, the work develops into meta-physics. Theories are often criticized for this, but it can't be helped. An explorer needs terrain to explore, and questions to ask, in order to stay in the game of theory. It is O.K. to use meta-physics as a tool to keep the engines of theory moving, as long as you don't rely on them. And since the origin of the universe is completely unknown, it becomes logical to ask even the deeper questions about nature, the universe,

God, and man. Using some of our known physics as framework, answers can be furnished with a reasonable amount of physics, yet the end result is usually a colorful weave of mythology and logic. Good physics can run parallel to the story telling, and evolve into logistical interpretations, which may also evolve into mathematical interpretations later on. This is exactly what happens in my theory workbook, and so the most interesting physics occurs early on before going into deep nature. Excerpts from the earlier chapters are worth reading, and can serve as a reference for some of the technical information presented so far. The chapter on gravity is a good study, as well is the last chapter of this book, which is based on the most recent edition of my theory workbook.

We are here to learn modern theorem, and to learn where it came from, and how it was figured out. All of this information will be here in this book. What will not be in this book is all of the theory that it took to get there. We don't need to study the philosophical, spiritual, or religious interpretations in the theory workbook, because they are not included in the mathematical language of a modern theorem. Mathematical interpretation can only be based on logistical interpretation, and for that you need physics. And you need to know the physics, if you are going to study modern theorem. Even though you may know a lot more now than I ever did when I first started out, it can still be a worthwhile experience to enter the mind of a theorist, and take the journey as I did, to the earliest moments of time. It is there that the question of the first energy and the first energy scenario awaits us.

Chapter Four- Review of the
Theory Workbook

EARLIER CHAPTERS (EXCERPTS)

What does it mean that the universe is expanding? Does it mean that there was a big bang to explain the expansion? The answer is yes. Science has been investigating this theory of a big bang for seventy years. The theory contains volumes of information as to the precise age of the opaque fireball which lasted ten-thousand tears, to the age of the fading explosion-expansion energy which continued progressing for another quarter of a million years. The elements fashioned during this time can be traced to the present to explain the chemical make-up of the universe. Indeed, the carbon based life on Earth can be pinpointed in time through examination of the cosmological record. What is known as the Planck moment is the smallest measurement of time we can go to the very beginning of the big bang. Science stops there. But from there, the symmetry of the big bang can be understood. The details would be fascinating to someone who understands chemistry or astrophysics. The overall simplicity of the perfection in the big bang is much easier to comprehend. In short, science believes the big bang started out as an equilibrium of sorts, where the four forces of nature, electromagnetism, gravity, the strong and the weak nuclear forces were all on an equal basis. They have since developed differently. The universe would have developed into pure radiation, had it not been for the existence of one extra baryon of matter per billion matter/antimatter pairs. Matter/antimatter pairs are radiation. Protons and neutrons are considered part of the baryon family of matter. The extra matter that came out of the big bang eventually gave us the planets and the stars. The formation of atoms depended on an exact mathematical prescription without room for variance during the development of the universe to assure a successful universe. Each of the four forces of nature have come to be understood in terms of their unique contributions to the formation of a successful universe. The weak nuclear force gave us atoms, and atoms gave us the elements.

"Time" as we perceive it is actually mass/energy with momentum, such as the start of the big bang. Time began with the big bang. Time should be redefined as space-time, a product of the big bang. Space-time has a certain momentum. The rate of this momentum gives us "the universe of space" we enjoy. The universe can be defined as mass/energy having a certain amount of expanding momentum. This would also be the true definition of "time". Time is simply mass/energy in motion. The energy equation is also a time equation, a reference for anything that moves.

Was there an implosion within the explosion, so that the big bang "out" was also the big bang "in"? This brings us to the "polarized big bang theory", a theory which has as its premise the conditions of a positive and negative universe being fashioned in the big bang, so that a matter universe is created as part and parcel of the simultaneous creation of an antimatter universe, giving us the effects of gravity, and allowing the matter universe not to run away with itself at escape velocity. Why didn't the antimatter universe run away with itself and suck us in, and why is it running at such a smooth fixed rate? Why is our universe of space-time in a state of equilibrium with gravity?

What is the antimatter universe "made of", and where did it go? What is the bonding mechanism between these two states of matter and antimatter? Is there such a thing as a "superposition of states" where both are compatible? What is the superposition? The superposition is where matter becomes so small, that it literally is between existence and non-existence. It is the threshold where both matter and antimatter can share the same states of existence. It is a medium. We know about this medium as a direct result of cosmological research. The evidence of virtual particles in our universe gives us this clue about the superposition of states. Virtual particles appear from out of nowhere and quickly annihilate. We know of their existence by the way they affect other particles. The fact that they come into existence at all gives us hope that we can explain how to get something from nothing, which is exactly what might have happened in the beginning of time.

The most prominent scientists of our day are admitting that they believe in a supreme being. It is the only logical explanation for the perfection in the equation of our universe. Nobel Prize winning Francis Crick, foremost scientist and author in the field of genetics, can not find a logical explanation for the formation of DNA. He believes the origin of the first life on our planet can only be attributed to an outside source, such as God, as the only logical explanation.

In school we were taught the difference of opinion between the creationists and the evolutionists. Things were simple back then. There were two sides to the question and only one could be right. Did it ever occur to anyone that perhaps both ideas were compatible, given a proper context?

The creationists believe in a higher power which is responsible for the existence of man. They believe that man was fashioned by God from the dust, and that the Earth too was created by God. The evolutionists are much more practical minded. Man simply evolved as a process of natural selection. There is much evidence to support this claim. Keep in mind that any explanation of the workings of the big bang itself would have to contain two schools of thought; one to satisfy the evolutionists and one to satisfy the creationists; one model without God, and one model with God.

Oh where oh where has the antimatter universe gone, oh where oh where can it be?

We have a pretty good idea of what was in the big bang as far as forces and particles from the very moment of the explosion to the cooling period and up to now. What we don't know is how it happened. We know what happened but not how. We know almost precisely what happened. We can say that the four forces, electromagnetism, the nuclear strong and the weak forces, and gravity, were all present in the big bang and on equal terms with each other at that moment. This is a very important piece of the puzzle. Another important piece is the fact that one extra baryon (proton/neutron family) came out of a billion matter/antimatter pairs, which gave us the planets and the stars. So now we know that the universe started out in an equilibrium of some kind, and that there was an imbalance in the laws of nature, which lucky for us, provided the universe with something besides radiation.

What is quantum electro dynamics, or QED? And what is meant by the "birth seed"? Birth seed- this unit of measurement would be so small, that an atom would be one hundred million times bigger; this is the seed from which our entire universe has grown; it is infinitesimally small, as in "quantum". Quantum electro dynamics- deals with the mathematical description of particle interaction and how the electromagnetic force acts between the charged particles; imagine being able to see the behavior of a collection of electrons for instance. Their "interaction" would constitute dynamics. Physicists are able to describe even more detail about these types of interactions on the quantum level using sets of equations.

If we are going to try and envision some kind of birth seed, we are going to have to envision some kind of interaction involving quantum electro dynamics as an exercise in logic if nothing else. Many tries have been made to find a logical framework for this phenomenon. The obvious trick is to get something from nothing, as we naturally assume that there was a vacuum of open spatial dimension to begin with, which is the closest thing to "nothing" we can think of. Upon examination of the record, we find a startling scientific achievement made by a physicist named Dirac, who discovered the existence of "virtual pairs" or particles of matter/antimatter that appear and quickly annihilate in a vacuum. We are fascinated to learn that these particles can be electron/positron pairs, and that as virtuals, they exist in a state of superposition, in between existence and non-existence. For some unexplained reason they travel in droves in our universe, appearing from nowhere, and annihilating in bursts of radiation. Most of our knowledge about them is attributed to the way they affect other particles.

Now that we know we can get something from nothing, our chances are improving. We now have a framework! Let's begin. Let's imagine a carefully worked out "QED" of two sets of virtual pairs. They decide to pop into the picture of open spatial dimension just long enough for us to get a good glimpse. We see that they are perfectly aligned in a single row across only inverted, with the electrons on the outside and the positrons on the inside. So there are four particles across in a paragene formation, like four zeros, 0 0 0 0. (Paragene is a new word meaning side by side positive and negative values. It is the root word for a very important word in modern theorem, "paragenesis").

We can't help but notice that the particles are not doing what they're supposed to be doing, and that's annihilate. Perhaps there is something different about this environment from the environment of our vacuum on Earth. The first thing that comes to mind is "gravity". There is no strong gravity around this QED yet, so maybe the virtuals will be spared their usual fate this time. We now have a classical mathematical description of a point particle- an electrically charged point particle has an infinite energy density in the electric field at the point. This simply means that in the openness of spatial dimension, our two electrons are emanating an electromagnetic field which is probably covering the entire spectrum of infinite spatial dimension, which in terms of time, would also constitute all past, present, and future. This can best be understood using Maxwell's Equations, which explain the behaviors of electromagnetism, and it's unusual ability to travel through time. The equations predict that an electromagnetic wave from an electron or signal source has two waves, a wave out and a wave in simultaneously, known as the retarded and advance waves. They are so fast in theory as to be in two places at once. Advance waves are never detected in our universe for reasons we will discover later on. It will suffice to say that our universe is believed to be finite in size, and that electromagnetic waves require the absence of boundary conditions before advance waves can be detected. So far we have learned about the behavior of electromagnetism when associated with the electron. There is a relationship between the two called infinite energy density, and the scope is the entire spectrum of open spatial dimension, so that the early universe was just one big electric field.

Everyone knows that likes repel and opposites attract. When two like particles touch they spark a fusion of energy. They overcome the electric repulsion force. We are going to need a fusion of energy in our QED...the zeros are no longer uniform across as before, spaced equally apart. The middle two are moving closer together, and if they touch, there will be a positive charge of light, a fusion of energy, such as 0 00 0. When the fusion is over, it will look like this, 0 0 0, showing only three particles, because the fusion becomes an infusion, and the positrons merge to become one particle.

And the light traveled out and away from the source, so,

how could it get back? The answer is the same way electromagnetic waves get back to their signal source, by becoming a continuum.

Meanwhile, the light was about to make a return entrance from behind, or under, or from within, or somewhere, and it was really moving…

We've addressed the behaviors of light and electromagnetism to illustrate their effectiveness in achieving "continuum" status. We simply use a circle to compare to a continuum as a means of explaining the dynamics of the continuum and how something can travel out and around and come back. The geometry of a circle is the perfect tool for this definition of dynamics.

Mathematically the light phase constituted infinity without beginning or end and assured the "Q" of an unlimited power source. For the "Q" to be relative to itself it would have to know no boundary conditions. The reference point in the circle of time was a boundary condition. Relativity could not be relative to itself with the existence of the opening in the continuum, the window, the reference point, the evidence. The circle of time had to be closed.

Relativity is now the beginning of time, which means, relativity is the first energy.

If the goal of relativity was perfection and it somehow was an inherent property, then this domino effect was ready for its grand finale. The circle of time-light (energy!) was a circle with a reference point. The reference point was the center particle. Any reference point in a circle makes that circle imperfect. If I can detect the reference point, the circle is not relative to itself. If on the other hand the circle is a perfect geometric circle, then I won't be able to tell where the starting point was. This same principle when applied to the continuum means that there was no room in the world of relativity for error or waste. The reference point represented a boundary condition on "perfection" and could not be allowed to exist. The energies in the "Q" were sufficient to solve this problem, so that time could be relative to itself. This final act would be the resolution of the "Q", and the end of the domino effect of relativity.

Classical physics has reached a conclusion based on the study of quantum electro dynamics that the quantum forces are alike to the larger forces when in the context of the beginning of time. This is important to remember and lends some credibility to the possibility of the birth seed being fashioned in this way. We cannot for the sake of saying these are only quantum particles turn our heads away from the power of the "Q". The evolutionists now have a theory of the birth seed based on the domino effect of relativity, a theory devoid of a supreme being or the formation of consciousness during the action of the "Q".

A mathematical model for the superposition of states is simply infinite sets or sequences in one dimension, or, infinity=one.

$$\infty = 1$$

THE PLANCK MOMENT

(Some of the information in this chapter is out-dated. Before the discovery of reflex cube there was no way to describe the exact mechanics in the birth of the universe. However much of what is being said in the older theory is accurate to some degree. I wanted to include this chapter as a reference for study, and for comparison to the more advanced science of reflex cube coming up next.)

UNIVERSE BIRTH EQUATION Theory Workbook- Authors Note: The literary style of this chapter is a departure from the form presented throughout the book. Since the focus of this chapter is "technical theory", I found it necessary to write the chapter in the manner of a formal scientific article, with a suitable amount of objectivity in the analysis to satisfy even the most critical of scientists. This is gravity theory and the math to prove it. This theory explains one of the most puzzling questions mankind has ever wondered about. The answers offered here have to be bold and clear, and presented as the data, and so I have altered the literary style to accommodate the formality.

If gravity is an equation, then it may be part of a larger equation. To fully understand gravity, it may be necessary to work the equation in reverse. We are dealing with forces, not numbers in our equations.

Gravity can be explained if the Big Bang was both positive and negative. There was a Big Bang "out", as well as a Big Bang "in". In other words, there was an explosion and an implosion. The obvious questions are what caused the "vacuum" for the implosion, and why is the anti-matter universe, a result of the implosion, running at such a smooth fixed rate giving us gravity? (FIGURE A)

To understand the "polarized" Big Bang model further, imagine such a thing as a time-line. The time-line is a line which begins as a simple point. The point becomes a line moving in two directions. The Big Bang is the point beginning the time-line. The Big Bang, our matter universe, moves to the left along the time-line, which would represent the future. In other words, we are living in the Big Bangs future. The negative side of the Big Bang, or anti-matter universe, moves to the right along the time-line and it goes into the anti-future, or the past. In between these two universes is a superposition of states. The superposition is a state between existence and non-existence, compatible with both universes of positive and negative. The superposition lives in the pure "present". The time-line shows the future/present/past relationships of the forces, and the directions they took in the Big Bang, but it does not explain the answers for a split equation at the Planck moment. Here is where we must work the equation in reverse, and here is where we will begin to see the larger equation. An examination of the Planck moment will not only yield the information we need to satisfy the new Big Bang model, it will raise more questions in light of the larger equation. Therefore, the logistical flow of the gravity equation, the part we are concerned with, will have to remain within the strict limitations of the Planck moment. What is needed to explain the vacuum for the implosion is a missing force. It is conceivable that the explosion itself might have created the proper conditions for a vacuum at its center, but this is highly unlikely in view of the tremendous energies involved. For all of this energy to have symmetrical expression, it must have had something besides the seed of itself as a reference. The missing force could have provided an alternative reference. Exactly what qualifications must this missing force have to satisfy the new Big Bang model? These are the questions we hope to answer with "M and M theory", a theory I have developed to satisfy the new model of a positive/negative Big Bang and to explain the nature of gravity.

Without going into the dynamics of the energy released in the Big Bang, we know that the energy was symmetrical, and that the four forces of nature were on an equal basis following the Planck moment. Because we know about the equilibrium of the forces this early, there must have been a correlation between this equilibrium and the missing force. This is one of the qualifications the missing force must have. It must also be a strong enough force to provide the energy of the Big Bang "out" with a counter-reaction of equilibrium. Another qualification the missing force must have is that it has to meet certain criteria. Identifying the missing force has to satisfy each condition of the new Big Bang model, and answer both questions of how the implosion took effect, and why the anti-matter universe is running at such a smooth fixed rate. The solution can only be arrived at using the same set of rules. Now that the preliminary guidelines have been drawn, we can try and identify the main qualifications of the missing force. It must be a force stronger than the Big Bang itself, and it must be contained within the seed. "M and M theory" satisfies all of these conditions.

The missing force must be nothing less than a physical reference point for a continuum of energy. Although continuums are outside the realm of verification, they are certainly not mathematically impossible. There are several ways to approach the problem of explaining continuum dynamics, but the easiest way to understand

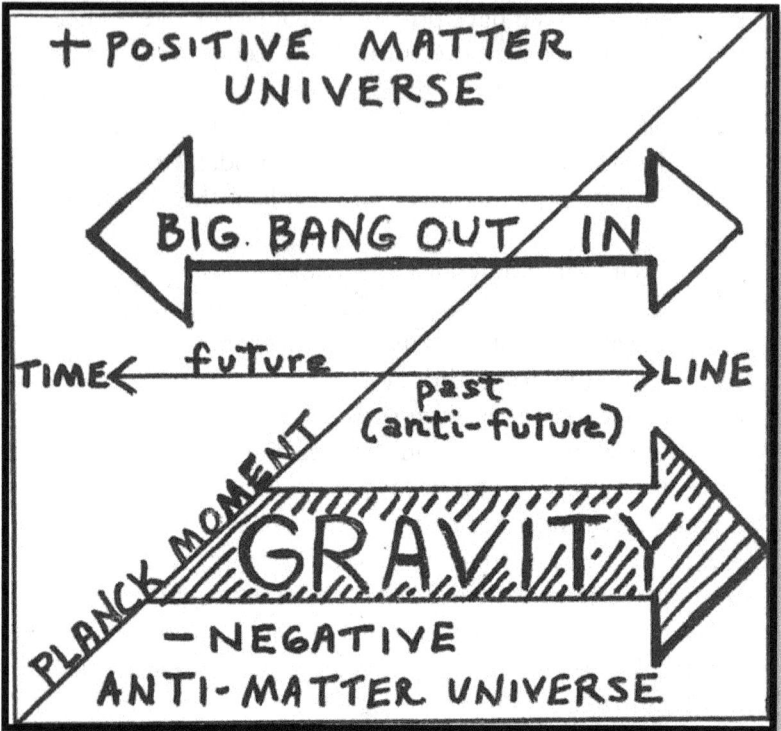

FIGURE A

The time-line brings into focus a working theory of gravity as being an artificial by-product of the Big Bang. According to a new model of the Big Bang, both a positive matter universe and a negative anti-matter universe were split in the equation at the Planck moment.

the continuum is the equation one=infinity. If there truly is such a thing as a continuum of energy, then it must have a reference point of resolution. In theory, the reference point is represented in the equation as "one". "M and M theory" brings into light the existence of a larger equation, but for now we will focus on the dynamics of the seed within the boundary conditions of the Planck moment. Revealed in the Planck moment is the sharing of two dimensions, the dimension of the continuum, and the dimension of space-time. (FIGURE B)

The separation of these two dimensions is called the "split of the inverse". In the Planck moment, the force of the blast registers in the continuum dimension. (This will be proved later in the theory). Perhaps the reference point of the continuum required compactification for synchronization, however this question leads into the dynamics associated with the singularity, and cannot be addressed here. Having identified the missing force as the reference point of a continuum is quite enough for now. The logistical flow of the forces can be illustrated to show the equilibrium needed in the new Big Bang model. The new dimension of space-time suddenly has a "gap" where the reference point of the continuum had been, providing a vacuum source. (FIGURE C)

Chronologically, this became evident directly outside of the boundary condition of the Planck moment, when it is believed all of the forces of nature were on an equal basis. The equilibrium of the forces at this moment plays a very important role in the development of this theory, as it suggests that the equilibrium of gravity must have been achieved almost instantaneously, which brings us to a complete definition of the gravity equation based on "relativity's all of time in suspension model of the inverse squared without resolution".

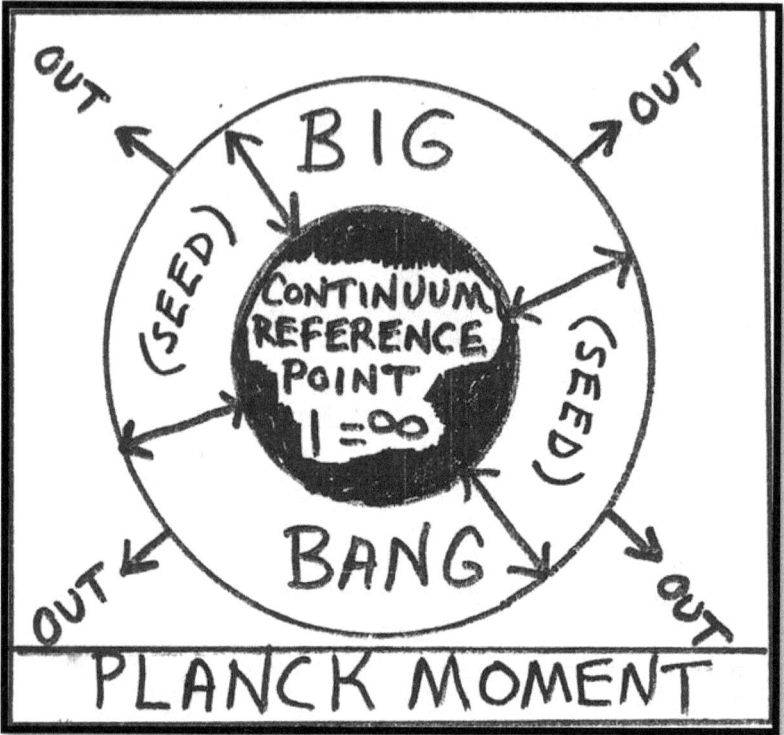

FIGURE B

The force of the blast must register in the continuum dimension in the Planck moment. This assures a uniform distribution of the energy in the Big Bang "out".

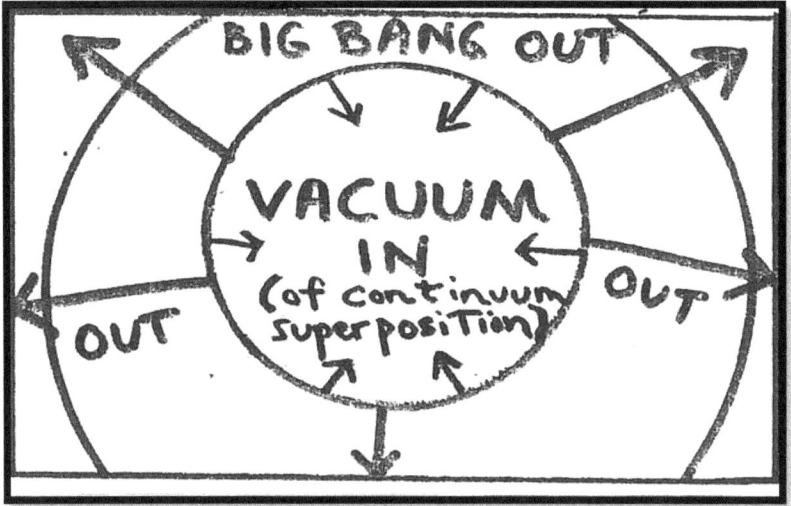

FIGURE C

At the split of the inverse and the separation of
dimensions a vacuum source is created for the implosion. Minus the reference point of the continuum, the
vacuum is filled with continuum superposition.

To understand this equation requires nothing more than basic knowledge in quantum physics. There are two ways matter can enter the superposition, 1) all matter has a telescoping range where it eventually meets the threshold of the superposition in between a state of existence and non-existence, and 2) under certain conditions a sudden displacement of matter can escape into the superposition as sequences, or copies, called "echo matter". The only boundary condition of the Planck moment which we are concerned about has been identified as the split of the inverse. The inverse itself should also be identified as the reference point of the continuum, as in one=infinity. Mathematically as well as in theory, the continuum must have a reference point, serving as both a source of its own energy and as a collector of its own energy. The split of the inverse clearly established the dimension of space-time as a dimension relative to itself, however there is still a connection of the superposition to it. This type of superposition is not a sudden displacement of just ordinary matter, as this matter has continuum momentum. It is a superposition of sequences having continuum momentum, and this accounts for the spontaneous reaction of equilibrium following the split of the inverse. Indeed the blast must have registered in the continuum dimension, allowing the superposition of the matter universe to get a head start, if by only a Planck moment. (FIGURE D).

The continuum superposition followed, and was not allowed to finalize a revolution. Had it been allowed, the Big Bang would have been nothing more than a Big Dud. The split of the inverse represents a boundary condition, where on one side of the boundary condition, the superposition of the matter universe begins with the Big Bang explosion "out", and on the other side, the superposition of the continuum begins with the vacuum implosion "in". The continuum superposition is not included in the Planck moment, and because of this, we have a rejected continuum giving us gravity. (FIGURE E)

Today the rejected continuum is trying to find its reference point to finalize a revolution but can't. Its starting place, the original "coordinates" of the vacuum, is carefully protected in the matter superposition of states. To find its original sequence, the continuum superposition must first penetrate our matter superposition, and since both states are alike, they repel. This equation, based on our physics, is called the inverse squared without resolution. The two stated of superposition are engaged in a squaring process at the boundary condition of our threshold of matter, where they enter a state of lesser reality, into the territory of inverted superficiality, also known as a state of suspension.

This equation of the inverse squared without resolution was realized in the moment following the Planck moment, the moment of independence for our universe, so that the forces of nature were all on an equal basis in the new dimension of space-time. Now that we know about the larger equation as evidenced by both dimensions sharing in the same Planck moment, we can no longer dismiss the singularity. Up until now, it has been mathematically correct to dismiss the singularity because the singularity resolves from a timeless state to another timeless state. It is merely a technicality that the resolution is from a state of non-existence to a state of existence, and this has been overlooked. Therefore it is incorrect both mathematically and in theory to dismiss the singularity, as the resolution into a timeless state is also the equalization of something else. The proof is in the gravity, and in the on-going equation.

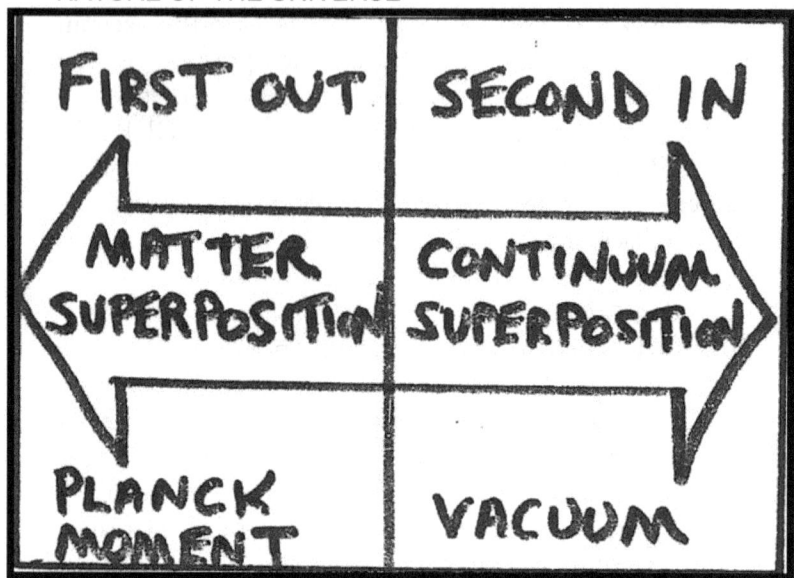

FIGURE D

Because the blast registered in the continuum dimension first, the matter universe got a head start. Consequently, the matter superposition went out first, before the continuum superposition went in second.

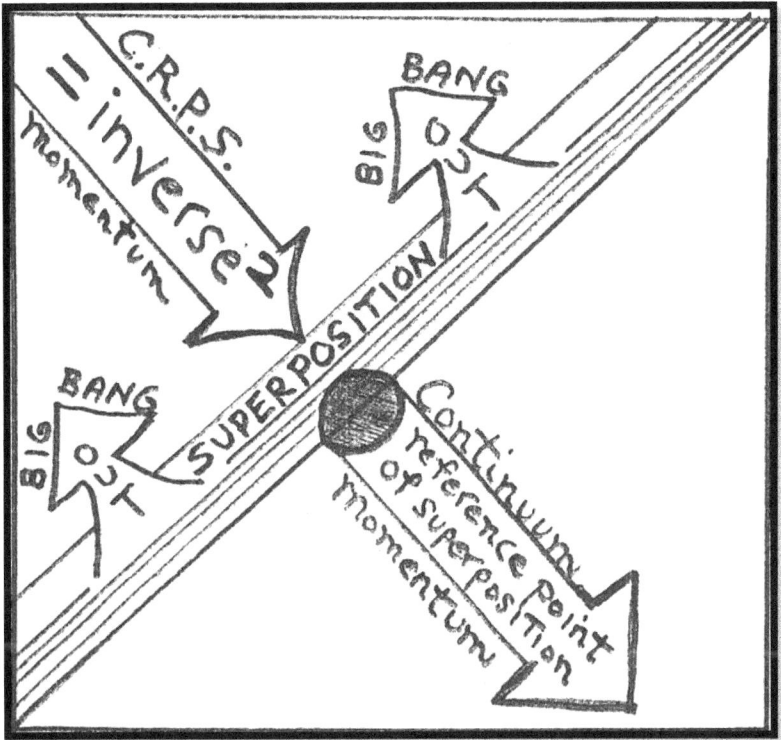

FIGURE E

"Relativity's all of time in suspension model of the inverse squared without resolution". A theory on the equilibrium of gravity may very well lead to a unified theory, even if this equilibrium requires the presence of a rejected continuum.

Mathematically there is an equalization superimposed on the resolution of the singularity, all because the Big Bang registered in the continuum dimension during the Planck moment. "M and M theory" stops here, reasoning that the singularity cannot be dismissed mathematically, as the singularity does not have true resolution.

Chapter Five-

The Math

REFLEX CUBE: THE LANGUAGE OF MODERN THEOREM

In our physical universe there is no such thing as transfinite, or the ability to count past all of the available numbers. To the best of our knowledge there is nothing faster than the speed of light, which is finite. C squared is also finite. No matter how fast you go, you can never outrun distance, if distance is infinite.

There are two very common denominators in our universe, distance (space), and motion (momentum). Everything is in motion and everything has room to move. According to classical physics, the equation "distance is always greater than momentum" is written as D>M. The opposite of this of course is M>D, and this is the equation of transfinite, assuming that spatial dimension or "space" is infinite.

Imagine if you will a train on tracks. The train represents "momentum" and the tracks represent "unlimited distance". According to our known physical laws, the train will never outrun the tracks, because the tracks are infinite, D>M. To understand continuum dynamics, we are going to have to imagine what might occur if M>D, and the train catches up to distance, and becomes "transfinite". At the point of catching up, the train would see itself and synchronize with itself. It would become 2 in 1 trains. The train would essentially fashion a self-replication, and it would have achieved "continuum momentum" in doing so, or M=D. This is called "continuum resolution".

There is only one example of self-replication that has a basis in reality, and that is DNA. (If you guessed DNA as the most unique thing in the universe earlier, then you guessed right). The ultimate secret of the continuum is that it makes copies of itself, but not just good copy like DNA. The continuum makes an absolutely perfect and authentic copy of itself at continuum resolution. The copy exists as a mirror image of itself in "reflection". The opposite of fusion is reflection. Reflection space is the domain of Hilbert space, or information space. The fusion is considered to be a non-dimensional without any space, and its opposite inversion of reflection is considered to be a one dimensional space, so that information can exist as information. Within this reflection space are going to be three copies of the continuum for a cube of information. This cube of information will reflect what is in the fusion.

As we said before, if nothing is all there is, then nothing is everything, a relativity continuum. This is only theoretically possible because a pair of opposites are sharing location. In our reality, opposites are opposed and are never allowed to share location. But in true nature, they are. This means that in nature, there exists a natural square of two things which are opposed. And because they are in containment in a fusion, they are also "one", or equal to each other. This is what makes a continuum so special. The opposites are opposed "in square", and they

have equivalency. We're going to find out that the opposites are also unopposed, and able to retain their independent identities as well. Now we should be quite familiar
with what type of information we can expect to find in the
reflection.

Let's get back to the example of the train. This is
how the math of continuum dynamics works: The train is
one train in reflection for 1=1, and the train is two trains
in reflection for 2=1, and the train is two trains in reflection for 2=2; one train in reflection for equivalency, two
trains in reflection for square, and two trains in reflection
as independents. These are the three synchronizations
which allow the continuum to be in three different places
at once. At continuum resolution, continuum momentum
becomes a multiple of synchronization called "positronic
synchronization". Positronic synchronization is perfectly
capable of maintaining a cube of reflection in one dimension of reflection for opposites which are the same,
opposed, and unopposed. It is better to think of the
continuum as an inverted continuum of synchronization,
rather than an open continuum as used in the example.
However, positronic synchronization is perfectly capable
of maintaining an actual cube of real dimension as well.

We know from the example that the equation
M=D, or "momentum equals distance", represents
continuum resolution. This equation is also referred to as
the "linear equivalent". Linear equivalency is easy to
understand using the axiom "the shortest distance between two points is a straight line". According to this
axiom, the original concept of space could be a straight
line of momentum representing all distance from zero to
infinity, in reflection to the opposite of no momentum or
distance at zero.

And now it's time for a review of what is referred

to as the "continuum dynamic". This is the principle of a fusion of opposites. You're not going to get a reflection unless you have a fusion of an opposite pair. This type of fusion can be read mathematically as 2 in 1. This is also the equation for square, or 1 square. 2 in 1 mathematical fusion has 3 in 1 reflection. This is called "reflex cube". Most opposites come in pairs. The square has two. Reflex cube is the double opposite inversion of the square for 1, 1 square, and the sum of the square in reflection.

(Lesson helper: The continuum dynamic is a two in one fusion of the opposites. Opposites are opposed. Two in one fusion is 1x1=1. There are two ones in one fusion. Each mathematical one represents one opposite of an opposite pair. The equation makes sense only if you view the opposites as one in the first place, and you have to, because they are in a fusion as one to begin with. Therefore, nature is a positive square for two in one, or 1x1=1. Nature, a positive square, is also a continuum, and a continuum has to contain the opposites. A positive square has two opposites, therefore a positive square is a unique system of double opposite inversion: A positive square is 1 squared. The opposite inversion of 1 squared is 1. Also, the opposite inversion of 1 squared is the sum of the square. Fusion has reflection. In reflection to the positive square is a cube of itself and the double opposite inversion. Therefore, two in one fusion has three in one reflection. The positive square is called "reflex". The cube in reflection is called "cube").

So the continuum dynamic has a cubed reflection of three continuums of synchronization:

Monoplex 1=1   positive   (opposites as one)
Reflex   2=1   positive square   (opposites together)
Multiplex 2=2 independent square (opposites separate)

REFLEX CUBE AND THE
NATURE OF THE UNIVERSE

The monoplex continuum is one dimensional and positive. The monoplex squared is reflex, or one squared. Reflex is the positive square for 1x1=1 and is two in one continuums for a centrosymmetric function of inclusion. The multiplex is two independent continuums as the sum of the square for a function of exclusion. The multiplex is the equation for equilibrium and even distribution. These three reflections of synchronization represent a cube of opposite inversion square called a "multiplex cube". Reflex cube refers to both the continuum dynamic and the multiplex cube.

The continuum dynamic is a positive square because there is no negative square in reflex cube. A negative square would of course be a positive anyway. Even though we allow a negative square to exist in our mathematics, nature does not allow this in the non-expressive. The continuum dynamic can be at rest as an inactive square in reflex cube (zero function), but it can also become an active square and set the multiplex cube into action (continuum function). This is when reflex cube becomes squared for the equation reflex cube squared. This is called "squaring the cube". This is when reflex cube goes into the expressive. This is when the fusion becomes squared for "infusion". This is when the precision based mechanics of positive and negative will begin to unfold in sequential order for the delivery of the correct "paragenesis", and the making of the space-time continuum. This action of nature is also the importation of a negative square in suspension. Reflex cube has to become squared in order to acquire the opposite of a permanent positive square. (Note: the term "multiplex" is used chiefly to describe what the cube is doing when it goes into action)

Now to explain the "UBE" (universe birth equation), all one needs to do is explain "positronic inversion". 13,700 million years ago, the linear equivalent went from M=D to M>D, and the multiplex squared into the reflection of what we now perceive as open spatial dimension. This was primary inversion, 3 to 1 centrosymmetric resolution of cubed reflection to one "positron", for a complete multiplex positive with resolution. Secondary inversion is the negative multiplex in reflection to the positron as an electron in centrosymmetric position going to cube or 1 to 3, with the negative square in reflex having to actuate its linear equivalent, but being blocked by the "positron". Reflex is a negative square in suspension, initiating the big bang and giving us gravity for a non-resolute cubed inversion, but with an even distribution of matter and anti-matter for a positive multiplex in resolution and a negative multiplex in non-resolution in an equilibrium of sum of square called "uniplex cube square".

Let me explain this in terms of simplification. Before the big bang there was no real space, only abstract space. We have a working theory as to what is in this abstract space. It has to be something which can build a real 3-dimensional space with height, width, and length in a 3 in 1 superposition as one space. The theory is that a cube of fusion in reflection can do this when it becomes squared, and squares in or out for what is called infusion. Closed space goes to its opposite inversion of open space instantaneously, and since this is a double inversion, macroscopic space without form goes to the next inversion of microscopic space with form, the first energy. This is the equation infinity=one and time hasn't even begun yet. We know that space has to come first. The rest is easy. When the negative starts to unfold, it goes in reverse order. Positive space doesn't have to register its linear equivalent, but negative gravity does, and so on. My advice is to keep on reading through the technical descriptions until you get it. All of the information you need to understand it is in the book.

NOTE: Primary inversion is complete when the positron is established in the monoplex. In secondary inversion, the electron begins in the monoplex as a double negative, for a positive, and has equivalency with the positron. Then in reflex, the electron separates, with one negative end remaining, and the other having to register its linear equivalent for square. In the meanwhile, matter and anti-matter separate. It only takes a billionth of a second for the other negative end to register its linear equivalent, but instead squares with matter. This is enough time for matter to get a head start without the resistance of gravity, so that a universe can be born. This gives the big bang a little extra "oomph" at the start, so that some of the virtual pairs can escape the fate of radiation. Astrophysicists have determined that 98% of the materials needed for a physical universe came out of the first seven minutes of the big bang.

The equation "uniplex cube square" is the equation for the birth of the universe, and since the universe is space dependent, the equation can also represent the space-time continuum.

Welcome to New Era Three Physics!!!

THE SINGULARITY SIMPLIFIED

Einstein put a postulate at the big bang singularity indicating that the singularity was a mathematical problem, and as we all know, mathematical problems require mathematical solutions. In his wildest dreams however, I doubt Einstein would have considered what today's physicists are contemplating about the singularity. The idea of both space and time coming out of the big bang has taken awhile to catch on, but now that it has, it becomes easier to understand the terms "infinite density" and "infinite dimension". The physicists are explaining it like this today. If you could roll back the film of the universe to the very beginning you would arrive at a point called the singularity. This is a point of infinite density, containing all of the matter in the universe, including all of spatial dimension as well, and so it is also a point of infinite dimension. Believing that space is infinite, most scientists can now relate to the term infinite dimension. A point can either be a moment in time, or it can represent a "mathematical point". We can now begin to visualize the term infinite density because we know about the power of black holes. We can imagine something so dense, that it can absorb everything around it. Something with infinite density would be able to swallow up everything including space, and then disappear. It would be the ultimate vacuum.

The term "singularity" means "single" or "one". What has eluded our numerology science into the question of one, and the nature of one in cosmology, is that "one" is also one squared. We were unaware of this application in the singularity, which is why the singularity has remained such a mystery.

We are now starting to realize the singularity as a vacuum of nothingness, and this gives us a more obvious mathematical application in the singularity. The best way to understand "nothing" mathematically is to describe it as "zero as an absolute". However, zero as an absolute is also everything as an absolute, and happens to be the only place in nature where opposites can have equivalency. Therefore, the original singularity is a relativity continuum, or a fusion of the opposites as "one". We would expect the original singularity to cancel out as zero, and it does. However, its reflection does not.

Theoretically, the singularity makes three copies of itself in one reflection. All three copies are indistinguishable from the original as perfect copies. In the reflection, the first copy represents the opposites as "one"; the second copy represents the opposites opposed in square for "one squared"; and the third copy represents the "independent square" (sum of the square), where the opposites are unopposed. These are actual copies of synchronization, and represent "information" for non-cancellation. Zero as an absolute is also represented for cancellation in the reflection as an independent square, so that the synchronization of information and no information are in perfect equilibrium.

So the singularity is nothing, and is not nothing. It is one, and it is not one. With the synchronization of continuum momentum, the opposites can have equivalency, be opposed, and unopposed, which is why the singularity likely has to exist in the pure "present". We'll never have to worry about duplicating continuum momentum, or disturbing the present, because it can't be done. If we could isolate the "present", or the present moment, then we would have something to worry about. We would then be able to effectively change or alter the past.

The singularity still remains at rest, and motionless in the present as reflex cube. When the square was activated, another reflection was made as opposite inversion copy to become the space-time continuum. This is when the singularity went into its future- a future with everything in motion in a universe with observers. Firstly, a reflection of three-dimensional open space has to be created instantaneously for the singularity to square into. Therefore, space has to exist in the pure present; where past and future have equivalency as one in the singularity. Time is heavier and slower, and exists as a universe in finite suspension, where the past and the future are opposed, in square, as gravity and energy. Observers have consciousness, and the illusion of the present moment, because the past and future are also unopposed, or separated. Therefore, we identify with the sum of the square to give us our reality, and are the last in the natural order of the singularity's enfoldment from instantaneous space, to a sequential of time, and finally to a reality of consciousness.

THE FIRST ENERGY REVISITED

There was a time about twenty years ago I thought any meaningful discussion about the first energy would have to wait and perhaps never happen. It was easy to figure out which type of energy was involved in the very beginning, based on what we know about radiation and positive and negative charge. But this only allows us to identify the best candidates for the first energy as having positive and negative charge, and still doesn't explain where the energy came from or how it got there. Once the big bang got started, the chain reaction continued on for many thousands of years, with the initial intensity taking that long to wind down. This makes the big bang one very long radiation event. There was a time when we could not explain where the water in the oceans came from, and how it all got there. It's a bit like asking the question of what was the first water drop, in hopes we might learn something about the origin of Earth's oceans.

I had spent a lot of time thinking about the question of the first energy during the course of writing my theory workbook, and I became ever so convinced that we would never know how to explain it. Yet soon after closing the theory workbook, I began to change my mind. This is because I discovered the idea of fusion and reflection. Later I was to learn that the idea had been around for some time. It was just never mentioned in any of the books I had read. I can still say that I discovered the idea for myself, and experienced the idea as a product of original thought. Imagine my delight when I took the next step and began to figure out infusion. With infusion, it logically follows that whether you go in or out of a fusion, it's going to take you to the same place. So the idea of infusion became just as fascinating as the idea of fusion, because it suggested that in nature, the opposites of in and out could be the same.

When I began to theorize about the fusion, I immediately realized that there had to be something else besides the fusion in nature. There had to be something else which would allow us as observers to observe and ask questions. So I tried to figure out what the opposite of a fusion would be, and the only thing I could think of was a reflection. I reasoned that if something isn't there, it can't possibly have any reflection. But since this is nature, and a place where opposites can be the same, I started to wonder. Maybe absolute nothingness can have a reflection of itself. Perhaps a reflection is a place where a fusion can do things it normally wouldn't do. At least the idea of a reflection would allow us as observers to theoretically observe and study the fusion indirectly, even though technically the fusion is a non-observable.

It wasn't long before I realized that infusion could be a fusion when squared, and soon I was on the way to discover the basic infrastructure of reflex cube, although I didn't have a name for it then. But I still had more questions than answers about the mechanics, and I knew that I was going to have to understand the mechanics of the system before the proper mathematics could be applied to describe them. So I had to figure out how a fusion could produce a reflection. I soon realized that the fusion was a continuum, and that continuums can make copies of themselves in a reflection. I had discovered the science of continuum resolution. But how could a fusion make a genuine copy of itself in an opposite inversion of reflection? It stands to reason that the copy of itself would be an opposite inversion copy of itself, like the reflection of something in a mirror. The answer would prove difficult, yet it was so simple. Ask yourself, what is the opposite inversion of fusion? You say reflection. I say, then how do you get a genuine copy of the fusion into the reflection, without the image being inverted. The answer is, the fusion, whether inverted or not, is the same, because in and out are the same. Therefore, the only thing in existence with the ability to produce "identicals" is the fusion. This is the basis for the mechanics of the monoplex continuum, where the opposites of fusion and reflection are one, in reflection. And now you can have two genuine copies of fusion opposed, in square, for a reflex continuum, in reflection, and again as two independent continuums unopposed for the sum of the square, in reflection. So these fusions are all identical participants in a cube of reflection, with each reflection having a different value. That's the beauty of nature, because now you have a multiplex cube in reflection to the original fusion.

Of course, it took awhile to understand all of this, and there was a learning curve. I was much more familiar with the mechanics of the system in certain areas, and more familiar with the mathematics of the system in others. You can't even prove that there is a system, unless you can describe all of the relationships of the system in a precise mathematical language. You can explain some of the principles of the mechanics, and some of the mathematics, but you will not be able to communicate it as a system without the rest of the information. This is similar to when an archeologist finds a large stone with ancient symbols carved into it, and the symbols are unfamiliar. He will spend a very long time trying to determine the meaning of the symbols, and whether the symbols represent a form of communication or not.

Having knowledge of the principles of reflex cube and some of the math is not enough to describe the dynamics of the first energy. That question was the one important question in modern cosmology which had eluded me. It is ironic that the term "positronic synchronization" was already in place long before I could explain the mechanics of the positron as the first energy. But eventually, I was able to do so. I used the rational of the super, superposition, and position, to come up with a scenario I called "positronic inversion". This would describe the double inversion from the super, to the superposition and position, and outline the introduction of space and time into reality. First there would be primary inversion for space as the superposition, and then secondary inversion for time as the position. The astrophysicists are going to be delighted with the science of positronic inversion for years to come. I will now attempt to describe primary and secondary inversion in even more detail.

I think the best way to learn about the first energy is to let the positron speak for itself.

"Hi! I'm the first positron and I am energy. I am also very special. I am the reference point for a continuum. I just arrived here at the centrosymmetric location of all of space as the opposite inversion of space, and I am in continuum resolution. I also have positive charge, unlike space, which is neutral. I am three-dimensional space with form. I am not macroscopic like space. I am microscopic. I am not in superposition. I have position. And I am not alone. My counterpart the electron will be here shortly".

The positron has just described itself in primary inversion. Time hasn't begun yet. Space is instantaneous and so is the positron in resolution. Infinity as a cube of space has the extraordinary ability to mathematically calculate its center. I'm sure the astrophysicists would be interested in the equations I have developed to show how this is done. In my theory workbook, I make mention of the first energy as being nothing less than a physical reference point for a continuum. The passage can be found in the chapter on the Planck moment right here in this book. And so I was on the right track, twenty years ago.

Let's remember what we're really talking about here. We're talking about infusion. Reflex cube is a cube of fusion and has just been squared for infusion. The infusion is going to produce a space-time continuum: a continuum without separation, and a continuum with separation. This is going to be a double inversion from "space", to "time", with time in a double inversion from positive to negative. Closed space goes to its opposite of open space instantaneously, and there is a positron in centrosymmetric resolution. The positron represents a reference point for a continuum and is one of two reference points. What will naturally follow the positron is the opposite inversion of the positron, the electron. The reference points for a continuum with separation, time, are being established. Space, a continuum without separation has already been established, and is a model of perfect symmetry without form. It stands to reason that the positron is also a model of perfect symmetry with form. Therefore, the positron may be the only thing in existence ever to have perfect geometric symmetry. When the positron said it was special, it really meant it.

This almost concludes primary inversion. Primary inversion is a 3 to 1 centrosymmetric resolution of cubed reflection to one "positron", for a complete multiplex positive with resolution. The double inversion of three-dimensional open space with the opposite inversion of a positron in resolution represents a double reflection. In reflection to the positron in resolution is the opposite inversion of itself, the electron, in a double reflection. The double electron is sharing in the moment as a double negative, for a positive, so that primary inversion and secondary inversion can synchronize in the negative monoplex continuum. Both continuum reference points for positive and negative are now synchronized in the monoplex continuum and time hasn't even begun yet. The next step in secondary inversion is negative reflex, and this time, with negative inversion, time is ready to begin. Let's ask the double electron to describe what it plans to do.

"We are no longer synchronized as one in the monoplex. We are going to be synchronized as one squared in reflex, which means we are going to have to separate. The way things are now, we are so much alike, that likes repel and we won't be able to square. One of us has to exit, and return as inversion copy to achieve the proper configuration for reflex. It will only take a nano-second to register the linear equivalent for inversion copy, and that's because of the rule for in and out are the same. Once out, the electron will find its way back into three-dimensional space from the opposite direction almost immediately, and resolve at its starting place to join the other in square. That's the plan anyway. Oops I forgot. Primary inversion and secondary inversion are still synchronized…did you just hear something"?

That was the electron's cue to conclude the interview and return to its starting place as inversion copy. The big bang has already begun. That's what happens when positive and negative charge are left alone together. He probably thought positive and negative charge would wait for him until he returned, or that maybe the positron would step out of the way for him when he got back. He wasn't counting on the fact that energy was now going to be in the way. Anyway, the electron doesn't get to resolve, and instead has to oppose energy, which is a good thing. Had he been allowed to resolve, the resulting double negative in square would have produced a positive, and there would be no gravity. Speaking of which, is why negative reflex is not allowed in the first place to unfold as a square. It can't. It is a double negative in the negative monoplex continuum and has to separate first in order to be opposed in reflex. The separation marks the initiation of the big bang, leaving the other electron to square to itself in the opposite inverse direction. Real space has already been provided for the electron to achieve linear equivalency in three-dimensions, and in a nano-second, it almost resolves to itself again for square. In that nano-second, the electron exists as pure momentum, creating a vacuum in its wake, and is still today trying to carry its forward momentum as far as possible. We all know the rest of the story, but here is the technical description of secondary inversion again anyway. When this makes perfect sense, then you are ready to teach reflex cube.

Secondary inversion is the negative multiplex in reflection to the positron as an electron in centrosymmetric position going to cube or 1 to 3, with the negative square in reflex having to actuate its linear equivalent, but being blocked by the "positron". Reflex is a negative square in suspension, initiating the big bang and giving us gravity for a non-resolute cubed inversion, but with an even distribution of matter and anti-matter for a positive multiplex in resolution and a negative multiplex in non-resolution in an equilibrium of sum of square called "uniplex cube square".

Welcome to the space-time continuum!!!

,

REFLEX CUBE AND THE
NATURE OF THE UNIVERSE

www.ingramcontent.com/pod-product-compliance
Lightning Source LLC
Chambersburg PA
CBHW070409200326
41518CB00011B/2123